LECTURES ON MATRICES

J. H. M. WEDDERBURN

DOVER PUBLICATIONS, INC.
Mineola, New York

DOVER PHOENIX EDITIONS

Bibliographical Note

This Dover edition, first published in 1964 and reissued in 2005, is an unabridged and unaltered republication of the work first published in 1934 by the American Mathematical Society as Volume XVII in their Colloquium Publications.

Library of Congress Cataloging-in-Publication Data

Wedderburn, J. H. M. (Joseph Henry Maclagan), 1882–1948.
Lectures on matrices / by J.H.M. Wedderburn.
p. cm. — (Dover phoenix editions)
Originally published: New York : American Mathematical Society, 1934, in series: Colloquim publications (American Mathematical Society) ; v. 19.
Includes bibliographical references and index.
ISBN 0-486-44167-9
1. Matrices. I. Title. II. Series.

QA188.W43 2005
512.9'434—dc22

2004059120

Manufactured in the United States of America
Dover Publications, Inc., 31 East 2nd Street, Mineola, N.Y. 11501

PREFACE

This book contains lectures on matrices given at Princeton University at various times since 1920. It was my intention to include full notes on the history of the subject, but this has proved impossible owing to circumstances beyond my control, and I have had to content myself with very brief notes (see Appendix I). A bibliography is given in Appendix II. In compiling it, especially for the period of the last twenty-five years, there was considerable difficulty in deciding whether to include certain papers which, if they had occurred earlier, would probably have found a place there. In the main, I have not included articles which do not use matrices as an algebraic calculus, or whose interest lies in some other part of mathematics rather than in the theory of matrices; but consistency in this has probably not been attained.

Since these lectures have been prepared over a somewhat lengthy period of time, they owe much to the criticism of many friends. In particular, Professor A. A. Albert and Dr. J. L. Dorroh read most of the MS making many suggestions, and the former gave material help in the preparation of the later sections of Chapter X.

J. H. M. WEDDERBURN.

Princeton, N. J.,
July 20, 1934.

CONTENTS

Chapter V

COMPOUND MATRICES

Chapter VI

SYMMETRIC, SKEW, AND HERMITIAN MATRICES

Chapter VII

COMMUTATIVE MATRICES

Chapter VIII

FUNCTIONS OF MATRICES

CHAPTER IX

THE AUTOMORPHIC TRANSFORMATION OF A BILINEAR FORM

CHAPTER X

LINEAR ASSOCIATIVE ALGEBRAS

APPENDIX I

APPENDIX II

CHAPTER I

MATRICES AND VECTORS

1.01 Linear transformations and vectors. In a set of linear equations

$$\begin{aligned}
\eta'_1 &= a_{11}\eta_1 + a_{12}\eta_2 + \cdots + a_{1n}\eta_n \\
\eta'_2 &= a_{21}\eta_1 + a_{22}\eta_2 + \cdots + a_{2n}\eta_n \\
&\cdots\cdots\cdots\cdots\cdots\cdots\cdots\cdots \\
\eta'_n &= a_{n1}\eta_1 + a_{n2}\eta_2 + \cdots + a_{nn}\eta_n
\end{aligned}$$

or

$$\eta'_i = \sum_{j=1}^{n} a_{ij}\eta_j \qquad (i = 1, 2, \cdots, n) \tag{1}$$

the quantities $\eta_1, \eta_2, \cdots, \eta_n$ may be regarded as the coordinates of a point P in n-space and the point $P'(\eta'_1, \eta'_2, \cdots, \eta'_n)$ is then said to be derived from P by the *linear homogeneous transformation* (1). Or, in place of regarding the η's as the coordinates of a point we may look on them as the components of a vector y and consider (1) as defining an operation which transforms y into a new vector y'. We shall be concerned here with the properties of such transformations, sometimes considered abstractly as entities in themselves, and sometimes in conjunction with vectors.

To prevent misconceptions as to their meaning we shall now define a few terms which are probably already familiar to the reader. By a *scalar* or number we mean an element of the field in which all coefficients of transformations and vectors are supposed to lie; unless otherwise stated the reader may assume that a scalar is an ordinary number real or complex.

A *vector*[1] of order n is defined as a set of n scalars $(\xi_1, \xi_2, \cdots, \xi_n)$ given in a definite order. This set, regarded as a single entity, is denoted by a single symbol, say x, and we write

$$x = (\xi_1, \xi_2, \cdots, \xi_n).$$

The scalars $\xi_1, \xi_2, \cdots, \xi_n$ are called the *coordinates* or *components* of the vector.

If $y = (\eta_1, \eta_2, \cdots, \eta_n)$ is also a vector, we say that $x = y$ if, and only if, corresponding coordinates are equal, that is, $\xi_i = \eta_i$ $(i = 1, 2, \cdots, n)$. The vector

$$z = (\zeta_1, \zeta_2, \cdots, \zeta_n) = (\xi_1 + \eta_1, \xi_2 + \eta_2, \cdots, \xi_n + \eta_n)$$

is called the *sum* of x and y and is written $x + y$; it is easily seen that the operation of addition so defined is commutative and associative, and it has a unique inverse if we agree to write 0 for the vector $(0, 0, \cdots, 0)$.

[1] In chapter 5 we shall find it convenient to use the name *hypernumber* for the term vector which is then used in a more restricted sense, which, however, does not conflict with the use made of it here.

If ρ is a scalar, we shall write

$$\rho x = x\rho = (\rho\xi_1, \rho\xi_2, \cdots, \rho\xi_n).$$

This is the only kind of multiplication we shall use regularly in connection with vectors.

1.02 **Linear dependence.** In this section we shall express in terms of vectors the familiar notions of linear dependence.[2] If $x_1, x_2, \cdots, x_r$ are vectors and $\omega_1, \omega_2, \cdots, \omega_r$ scalars, any vector of the form

$$x = \omega_1 x_1 + \omega_2 x_2 + \cdots + \omega_r x_r \tag{2}$$

is said to be *linearly dependent* on $x_1, x_2, \cdots, x_r$; and these vectors are called linearly independent if an equation which is reducible to the form

$$0 = \omega_1 x_1 + \omega_2 x_2 + \cdots + \omega_r x_r$$

can only be true when each $\omega_i = 0$. Geometrically the r vectors determine an r-dimensional subspace of the original n-space and, if $x_1, x_2, \cdots, x_r$ are taken as the coordinate axes, $\omega_1, \omega_2, \cdots, \omega_r$ in (2) are the coordinates of x.

We shall call the totality of vectors x of the form (2) the *linear set* or *subspace* $(x_1, x_2, \cdots, x_r)$ and, when $x_1, x_2, \cdots, x_r$ are linearly independent, they are said to form a *basis* of the set. The number of elements in a basis of a set is called the *order* of the set.

Suppose now that $(x_1, x_2, \cdots, x_r)$, $(y_1, y_2, \cdots, y_s)$ are bases of the same linear set and assume $s \geq r$. Since the x's form a basis, each y can be expressed in the form

$$y_i = a_{i1}x_1 + a_{i2}x_2 + \cdots + a_{ir}x_r \qquad (i = 1, 2, \cdots, s) \tag{3}$$

and, since the y's form a basis, we may set

$$x_i = b_{i1}y_1 + b_{i2}y_2 + \cdots + b_{is}y_s \qquad (i = 1, 2, \cdots, r)$$

and therefore from (3)

$$y_i = \sum_{j=1}^{r} a_{ij}x_j = \sum_{j=1}^{r} a_{ij} \sum_{k=1}^{s} b_{jk}y_k = \sum_{k=1}^{s} c_{ik}y_k, \tag{4}$$

where $c_{ik} = \sum_{j=1}^{r} a_{ij}b_{jk}$, which may also be written

$$c_{ik} = \sum_{j=1}^{s} a_{ij}b_{jk} \qquad (i = 1, 2, \cdots, s) \tag{5}$$

if we agree to set $a_{ij} = 0$ when $j > r$. Since the y's are linearly independent, (4) can only hold true if $c_{ii} = 1$, $c_{ik} = 0$ $(i \neq k)$ so that the determinant

[2] See for instance Bôcher, *Introduction to Higher Algebra*, p. 34.

$|c_{ik}| = 1$. But from the rule for forming the product of two determinants it follows from (5) that $|c_{ik}| = |a_{ik}||b_{ik}|$ which implies (i) that $|a_{ik}| \neq 0$ and (ii) that $r = s$, since otherwise $|a_{ik}|$ contains the column $a_{i,\, r+1}$ each element of which is 0. The order of a set is therefore independent of the basis chosen to represent it.

It follows readily from the theory of linear equations (or from §1.11 below) that, if $|a_{ij}| \neq 0$ in (3), then these equations can be solved for the x's in terms of the y's, so that the conditions established above are sufficient as well as necessary in order that the y's shall form a basis.

If e_i denotes the vector whose ith coordinate is 1 and whose other coordinates are 0, we see immediately that we may write

$$x = \xi_1 e_1 + \xi_2 e_2 + \cdots + \xi_n e_n$$

in place of $x = (\xi_1, \xi_2, \cdots, \xi_n)$. Hence $e_1, e_2, \cdots, e_n$ form a basis of our n-space. We shall call this the *fundamental basis* and the individual vectors e_i the fundamental *unit vectors*.

If $x_1, x_2, \cdots, x_r (r < n)$ is a basis of a subspace of order r, we can always find $n-r$ vectors $x_{r+1}, \cdots, x_n$ such that $x_1, x_2, \cdots, x_n$ is a basis of the fundamental space. For, if x_{r+1} is any vector not lying in $(x_1, x_2, \cdots, x_r)$, there cannot be any relation

$$\omega_1 x_1 + \omega_2 x_2 + \cdots + \omega_r x_r + \omega_{r+1} x_{r+1} = 0$$

in which $\omega_{r+1} \neq 0$ (in fact every ω must be 0) and hence the order of $(x_1, x_2, \cdots, x_r, x_{r+1})$ is $r + 1$. Since the order of $(e_1, e_2, \cdots, e_n)$ is n, a repetition of this process leads to a basis $x_1, x_2, \cdots, x_r, \cdots, x_n$ of order n after a finite number of steps; a suitably chosen e_i may be taken for x_{r+1}. The $(n-r)$-space $(x_{r+1}, \cdots, x_n)$ is said to be *complementary* to $(x_1, x_2, \cdots, x_r)$; it is of course not unique.

1.03 **Linear vector functions and matrices.** The set of linear equations given in §1.01, namely,

$$\eta_i' = \sum_{j=1}^{n} a_{ij}\eta_j \qquad (i = 1, 2, \cdots, n) \tag{6}$$

define the vector $y' = (\eta_1', \eta_2', \cdots, \eta_n')$ as a linear homogeneous function of the coordinates of $y = (\eta_1, \eta_2, \cdots, \eta_n)$ and in accordance with the usual functional notation it is natural to write $y' = A(y)$; it is usual to omit the brackets and we therefore set in place of (6)

$$y' = Ay.$$

The function or operator A when regarded as a single entity is called a matrix; it is completely determined, relatively to the fundamental basis, when the n^2 numbers a_{ij} are known, in much the same way as the vector y is determined by its coordinates. We call the a_{ij} the *coordinates* of A and write

$$A = \begin{Vmatrix} a_{11} & a_{12} & \cdots & a_{1n} \\ a_{21} & a_{22} & \cdots & a_{2n} \\ \cdot\ \cdot\ \cdot & \cdot & \cdots & \cdot \\ \cdot\ \cdot\ \cdot & \cdot & \cdots & \cdot \\ a_{n1} & a_{n2} & \cdots & a_{nn} \end{Vmatrix} \tag{7}$$

or, when convenient, $A = \| a_{ij} \|$. It should be noted that in a_{ij} the first suffix denotes the row in which the coordinate occurs while the second gives the column.

If $B = \| b_{ij} \|$ is a second matrix, $y'' = A(By)$ is a vector which is a linear vector homogeneous function of y, and from (6) we have

$$\eta_i'' = \sum_{p=1}^{n} a_{ip} \sum_{p=1}^{n} b_{pj}\eta_j = \sum_{j=1}^{n} d_{ij}\eta_j,$$

where

$$d_{ij} = \sum_{p=1}^{n} a_{ip}b_{pj}. \tag{8}$$

The matrix $D = \| d_{ij} \|$ is called the *product* of A into B and is written AB. The form of (8) should be carefully noted; in it each element of the ith row of A is multiplied into the corresponding element of the jth column of B and the terms so formed are added. Since the rows and columns are not interchangeable, AB is in general different from BA; for instance

$$\begin{Vmatrix} 1 & 0 \\ 2 & 1 \end{Vmatrix} \begin{Vmatrix} a & b \\ c & d \end{Vmatrix} = \begin{Vmatrix} a & b \\ 2a + c & 2b + d \end{Vmatrix}$$

$$\begin{Vmatrix} a & b \\ c & d \end{Vmatrix} \begin{Vmatrix} 1 & 0 \\ 2 & 1 \end{Vmatrix} = \begin{Vmatrix} a + 2b & b \\ c + 2d & d \end{Vmatrix}$$

The product defined by (8) is associative; for if $C = \| c_{ij} \|$, the element in the ith row and jth column of $(AB)C$ is

$$\sum_{q=1}^{n} \left(\sum_{p=1}^{n} a_{ip}b_{pq} \right) c_{qj} = \sum_{p=1}^{n} a_{ip} \left(\sum_{q=1}^{n} b_{pq}c_{qj} \right)$$

and the term on the right is the (i, j) coordinate of $A(BC)$.

If we add the vectors Ay and By, we get a vector whose ith coordinate is (cf. (6))

$$\eta_i' = \sum_{j=1}^{n} a_{ij}\eta_j + \sum_{j=1}^{n} b_{ij}\eta_j = \sum_{j=1}^{n} c_{ij}\eta_j$$

where $c_{ij} = a_{ij} + b_{ij}$. Hence $Ay + By$ may be written Cy where $C = \| c_{ij} \|$. We define C to be the *sum* of A and B and write $C = A + B$; two matrices are then added by adding corresponding coordinates just as in the case of vectors. It follows immediately from the definition of sum and product that

$$A + B = B + A, \qquad (A + B) + C = A + (B + C),$$

$$A(B + C) = AB + AC, \qquad (B + C)A = BA + CA,$$

$$A(x + y) = Ax + Ay,$$

A, B, C being any matrices and x, y vectors. Also, if k is a scalar and we set $y' = Ay$, $y'' = ky'$, then

$$y'' = ky' = kA(y) = A(ky)$$

or in terms of the coordinates

$$\eta_i'' = \sum_j k a_{ij} \eta_j.$$

Hence kA may be interpreted as the matrix derived from A by multiplying each coordinate of A by k.

On the analogy of the unit vectors e_i we now define the *fundamental unit matrices* e_{ij} $(i, j = 1, 2, \cdots, n)$. Here e_{ij} is the matrix whose coordinates are all 0 except the one in the ith row and jth column whose value is 1. Corresponding to the form $\Sigma \xi_i e_i$ for a vector we then have

(9)
$$A = \sum_{i,j=1}^{n} a_{ij} e_{ij}.$$

Also from the definition of multiplication in (8)

(10)
$$e_{ij} e_{jk} = e_{ik}, \qquad e_{ij} e_{pq} = 0, \qquad (j \neq p)$$

a set of relations which might have been made the basis of the definition of the product of two matrices. It should be noted that it follows from the definition of e_{ij} that

(11)
$$e_{ij} e_j = e_i, \qquad e_{ij} e_k = 0 \qquad (j \neq k),$$

(12)
$$Ae_k = \sum_{i,j} a_{ij} e_{ij} e_k = \sum_i a_{ik} e_i.$$

Hence the coordinates of Ae_k are the coordinates of A that lie in the kth column.

1.04 **Scalar matrices.** If k is a scalar, the matrix K defined by $Ky = ky$ is called a *scalar matrix*; from (1) it follows that, if $K = \| k_{ij} \|$, then $k_{ii} = k$ $(i = 1, 2, \cdots, n)$, $k_{ij} = 0$ $(i \neq j)$. The scalar matrix for which $k = 1$ is called the identity matrix of order n; it is commonly denoted by I but, for reasons explained below, we shall here usually denote it by 1, or by 1_n if it is desired to indicate the order. When written at length we have

$$1_n = \left\| \begin{matrix} 1 & & & & \\ & 1 & & & \\ & & \cdot & & \\ & & & \cdot & \\ & & & & \cdot \\ & & & & & 1 \end{matrix} \right\|, \quad K = \left\| \begin{matrix} k & & & & \\ & k & & & \\ & & \cdot & & \\ & & & \cdot & \\ & & & & \cdot \\ & & & & & k \end{matrix} \right\|$$

A convenient notation for the coordinates of the identity matrix was introduced by Kronecker: if δ_{ij} is the numerical function of the integers i, j defined by

$$(13) \qquad \delta_{ii} = 1, \qquad \delta_{ij} = 0 \qquad (i \neq j),$$

then $1_n = \| \delta_{ij} \|$. We shall use this Kronecker delta function in future without further comment.

THEOREM 1. *Every matrix is commutative with a scalar matrix.*

Let k be the scalar and $K = \| k_{ij} \| = \| k\delta_{ij} \|$ the corresponding matrix. If $A = \| a_{ij} \|$ is any matrix, then from the definition of multiplication

$$KA = \left\| \sum_p k_{ip} a_{pj} \right\| = \left\| \sum_p k\delta_{ip} a_{pj} \right\| = \| ka_{ij} \|$$

$$AK = \left\| \sum_p a_{ip} k_{pj} \right\| = \left\| \sum_p ka_{ip} \delta_{pj} \right\| = \| ka_{ij} \|$$

so that $AK = KA$.

If k and h are two scalars and K, H the corresponding scalar matrices, then $K + H$ and KH are the scalar matrices corresponding to $k + h$ and kh. Hence the *one-to-one* correspondence between scalars and scalar matrices is maintained under the operations of addition and multiplication, that is, the two sets are simply isomorphic with respect to these operations. So long therefore as we are concerned only with matrices of given order, there is no confusion introduced if we replace each scalar by its corresponding scalar matrix, just as in the theory of ordinary complex numbers, $(a, b) = a + bi$, the set of numbers of the form $(a, 0)$ is identified with the real continuum. We shall therefore as a rule denote $\| \delta_{ij} \|$ by 1 and $\| k\delta_{ij} \|$ by k.

1.05 Powers of a matrix; adjoint matrices. Positive integral powers of $A = \| a_{ij} \|$ are readily defined by induction; thus

$$A^2 = A \cdot A, \qquad A^3 = A \cdot A^2, \cdots, \qquad A^m = A \cdot A^{m-1}.$$

With this definition it is clear that $A^r A^s = A^{r+s}$ for any positive integers r, s. Negative powers, however, require more careful consideration.

Let the determinant formed from the array of coefficients of a matrix be denoted by

$$| A | = \text{det. } A$$

and let α_{qp} be the cofactor of a_{pq} in A, so that from the properties of determinants

(14) $$\sum_p a_{ip}\alpha_{pj} = | A | \delta_{ij} = \sum_p \alpha_{ip}a_{pj} \qquad (i, j = 1, 2, \cdots, n).$$

The matrix $\| \alpha_{ij} \|$ is called the *adjoint* of A and is denoted by adj A. In this notation (14) may be written

(15) $$A(\text{adj } A) = | A | = (\text{adj } A)A,$$

so that a matrix and its adjoint are commutative.

If $| A | \neq 0$, we define A^{-1} by

(16) $$A^{-1} = | A |^{-1} \text{adj } A.$$

Negative integral powers are then defined by $A^{-r} = (A^{-1})^r$; evidently $A^{-r} = (A^r)^{-1}$. We also set $A^0 = 1$, but it will appear later that a different interpretation must be given when $| A | = 0$. Since $AB \cdot B^{-1}A^{-1} = A \cdot BB^{-1} \cdot A^{-1} = AA^{-1} = 1$, the reciprocal of the product AB is

$$(AB)^{-1} = B^{-1}A^{-1}.$$

If A and B are matrices, the rule for multiplying determinants, when stated in our notation, becomes

$$| AB | = | A | | B |.$$

In particular, if $AB = 1$, then $| A | | B | = 1$; hence, if $| A | = 0$, there is no matrix B such that $AB = 1$ or $BA = 1$. The reader should notice that, if k is a scalar matrix of order n, then $| k | = k^n$.

If $A = 0$, A is said to be *singular*; if $A \neq 0$, A is *regular* or non-singular. When A is regular, A^{-1} is the only solution of $AX = 1$ or of $XA = 1$. For, if $AX = 1$, then

$$A^{-1} = A^{-1} \cdot 1 = A^{-1}AX = X.$$

If $AX = 0$, then either $X = 0$ or A is singular; for, if A^{-1} exists,

$$0 = A^{-1}Ax = X.$$

If $A^2 = A \neq 0$, then A is said to be *idempotent*; for example e_{11} and $\begin{Vmatrix} 4 & -2 \\ 6 & -3 \end{Vmatrix}$ are idempotent. A matrix a power of which is 0 is called *nilpotent*. If the lowest power of A which is 0 is A^r, r is called the *index* of A; for example, if $A = e_{12} + e_{23} + e_{34}$, then

$$A^2 = e_{13} + e_{24}, \qquad A^3 = e_{14}, \qquad A^4 = 0,$$

so that the index of A in this case is 4.

1.06 The transverse of a matrix. If $A = \| a_{ij} \|$, the matrix $\| a'_{ij} \|$ in which $a'_{ij} = a_{ji}$ is called the *transverse*[3] of A and is denoted by A'. For instance the transverse of

$$\left\| \begin{matrix} a_{11} & a_{12} & a_{13} \\ a_{21} & a_{22} & a_{23} \\ a_{31} & a_{32} & a_{33} \end{matrix} \right\| \quad \text{is} \quad \left\| \begin{matrix} a_{11} & a_{21} & a_{31} \\ a_{12} & a_{22} & a_{32} \\ a_{13} & a_{23} & a_{33} \end{matrix} \right\|$$

The transverse, then, is obtained by the interchange of corresponding rows and columns. It must be carefully noted that this definition is relative to a particular set of fundamental units and, if these are altered, the transverse must also be changed.

THEOREM 2. *The transverse of a sum is the sum of the transverses of the separate terms, and the transverse of a product is the product of the transverses of the separate factors in the reverse order.*

The proof of the first part of the theorem is immediate and is left to the reader. To prove the second it is sufficient to consider two factors. Let $A = \| a_{ij} \|$, $B = \| b_{ij} \|$, $C = AB = \| c_{ij} \|$ and, as above, set $a'_{ij} = a_{ji}$, $b'_{ij} = b_{ji}$, $c'_{ij} = c_{ji}$; then

$$c'_{ij} = c_{ji} = \sum_p a_{jp} b_{pi} = \sum_p b'_{ip} a'_{pj}$$

whence

$$(AB)' = C' = B'A'.$$

The proof for any number of factors follows by induction.

If $A = A'$, A is said to be *symmetric* and, if $A = -A'$, it is called *skew-symmetric* or *skew*. A scalar matrix k is symmetric and the transverse of kA is kA'.

THEOREM 3. *Every matrix can be expressed uniquely as the sum of a symmetric and a skew matrix.*

For if $A = B + C$, $B' = B$, $C' = -C$, then $A' = B' + C' = B - C$ and therefore

$$B = (A + A')/2, \qquad C = (A - A')/2.$$

Conversely $2A = (A + A') + (A - A')$ and $A + A'$ is symmetric, $A - A'$ skew.

[3] It is also called the transposed or conjugate of A. It is sometimes written $\breve{A}$.

1.07 **Bilinear forms.** A scalar bilinear form in two variable vectors, $x = \Sigma \xi_i e_i$, $y = \Sigma \eta_i e_i$, is a function of the form

$$A(x, y) = \sum_{i,j=1}^{n} a_{ij} \xi_i \eta_j. \tag{17}$$

There is therefore a one-to-one correspondence between such forms and matrices, $A = \| a_{ij} \|$ corresponding to $A(x, y)$. The special form for which $A = \| \delta_{ij} \| = 1$ is of very frequent occurrence and we shall denote it by S; it is convenient to omit the brackets and write simply

$$Sxy = \xi_1 \eta_1 + \xi_2 \eta_2 + \cdots + \xi_n \eta_n \tag{18}$$

and, because of the manner in which it appears in vector analysis, we shall call it the *scalar* of xy. Since S is symmetric, $Sxy = Syx$.

The function (17) can be conveniently expressed in terms of A and S; for we may write $A(x, y)$ in the form

$$A(x, y) = \sum_{i=1}^{n} \xi_i \left(\sum_{j=1}^{n} a_{ij} \eta_j \right) = SxAy.$$

It may also be written

$$\sum_{j=1}^{n} \left(\sum_{i=1}^{n} a_{ij} \xi_i \right) \eta_j = SA'xy = SyA'x;$$

hence

$$SxAy = SyA'x, \tag{19}$$

so that the form (17) is unaltered when x and y are interchanged if at the same time A is changed into A'. This gives another proof of Theorem 2. For

$$Sx(AB)'y = SyABx = SBxA'y = SxB'A'y,$$

which gives $(AB)' = B'A'$ since x and y are independent variables.

1.08 **Change of basis.** We shall now investigate more closely the effect of a change in the fundamental basis on the coordinates of a vector or matrix. If $f_1, f_2, \cdots, f_n$ is a basis of our n-space, we have seen (§1.02) that the f's are linearly independent. Let

$$f_i = \sum_{j=1}^{n} p_{ji} e_j = Pe_i \qquad (i = 1, 2, \cdots, n)$$
$$P = \| p_{ij} \|. \tag{20}$$

Since the f's form a basis, the e's are linearly expressible in terms of them, say

$$e_i = \sum_{j=1}^{n} q_{ji} f_j, \tag{21}$$

and, if $Q = \| q_{ij} \|$, this may be written

$$(22) \qquad e_i = \sum_j q_{ji} \sum_k p_{kj} e_k = PQe_i \qquad (i = 1, 2, \cdots, n).$$

Hence $PQ = 1$, which is only possible if $| P | \neq 0$, $Q = P^{-1}$.

Conversely, if $| P | \neq 0$, $Q = P^{-1}$, and $f_i = Pe_i$ as in (20), then (22) holds and therefore also (21), that is, the e's, and therefore also any vector x, are linearly expressible in terms of the f's. We have therefore the following theorem.

THEOREM 4. *If $f_i = Pe_i$ $(i = 1, 2, \cdots, n)$, the vectors f_i form a basis if, and only if, $| P | \neq 0$.*

If we have fewer than n vectors, say $f_1, f_2, \cdots, f_r$, we have seen in 1.02 that we can choose $f_{r+1}, \cdots, f_n$ so that $f_1, f_2, \cdots, f_n$ form a basis. Hence

THEOREM 5. *If $f_1, f_2, \cdots, f_r$ are linearly independent, there exists at least one non-singular matrix P such that $Pe_i = f_i$ $(i = 1, 2, \cdots, r)$.*

We shall now determine how the form Sxy, which was defined relatively to the fundamental basis, is altered by a change of basis. As above let

$$(23) \quad f_i = Pe_i, \qquad e_i = P^{-1}f_i = Qf_i, \qquad | P | \neq 0, \qquad (i = 1, 2, \cdots, n)$$

be a basis and

$$x = \Sigma \xi_i e_i = \Sigma \xi'_i f_i, \qquad y = \Sigma \eta_i e_i = \Sigma \eta'_i f_i$$

variable vectors; then from (23)

$$x = Q\Sigma \xi_i f_i = P\Sigma \xi'_i e_i, \qquad y = Q\Sigma \eta_i f_i = P\Sigma \eta'_i e_i$$

and

$$\Sigma \xi'_i e_i = P^{-1}x = Qx, \qquad \Sigma \eta'_i e_i = Qy.$$

Let us set temporarily $S_e xy$ for Sxy and also put $S_f xy = \Sigma \xi'_i \eta'_i$, the corresponding form with reference to the new basis; then

$$(24) \qquad \begin{aligned} S_f xy &= S_e QxQy = S_e xQ'Qy \\ S_e xy &= S_f PxPy. \end{aligned}$$

Consider now a matrix $A = \| a_{ij} \|$ defined relatively to the fundamental basis and let A_1 be the matrix which has the same coordinates when expressed in terms of the new basis as A has in the old. From the definition of A and from $\xi_j = S_e e_j x$ we have

$$Ax = \sum_{i,j} a_{ij} \xi_j e_i = \sum_{i,j} a_{ij} e_i S_e e_j x$$

and hence

$$(25)\qquad \begin{aligned} A_1x &= \Sigma a_{ij}\xi'_j f_i = \Sigma a_{ij} f_i S_f f_j x = \Sigma a_{ij} Q^{-1} e_i S_e Q f_j Q x \\ &= Q^{-1}\Sigma a_{ij} e_i S_e e_j Q x = Q^{-1}AQx. \end{aligned}$$

We have therefore, remembering that $Q = P^{-1}$,

THEOREM 6. *If $f_i = Pe_i$ $(i = 1, 2, \cdots, n)$ is a basis and A any matrix, the matrix PAP^{-1} has the same coordinates when expressed in terms of this basis as A has in terms of the fundamental basis.*

The matrix $Q^{-1}AQ$ is said to be *similar* to A and to be the *transform* of A by Q. Obviously the transform of a product (sum) is the product (sum) of the transforms of the individual factors (terms) with the order unaltered. For instance $Q^{-1}ABQ = Q^{-1}AQ\cdot Q^{-1}BQ$.

Theorem 6 gives the transformation of the matric units e_{ij} defined in §1.03 which corresponds to the vector transformation (23); the result is that, if f_{ij} is the unit in the new system corresponding to e_{ij}, then

$$f_{ij} = Pe_{ij}P^{-1}$$

which is readily verified by setting

$$A = e_{ij} = e_i S_e e_j(\ \ ), \qquad A_1 = f_{ij} = f_i S_f f_j(\ \ )$$

in (25). The effect of the change of basis on the form of the transverse is found as follows. Let A^* be defined by

$$S_f x A y = S_f y A^* x;$$

then

$$\begin{aligned} S_f y A^* x = S_f x A y &= S_e Q x Q A y = S_e x Q' Q A y = S_e Q y (Q') A' Q' Q x \\ &= S_f y (Q'Q) A' Q' Q x. \end{aligned}$$

Hence

$$(26)\qquad A^* = (Q'Q)A'Q'Q.$$

1.09 Reciprocal and orthogonal bases. With the same notation as in the previous section we have $S_f f_i f_j = 0$ $(i \neq j)$, $S_f f_i f_i = 1$. Hence

$$\delta_{ij} = S_f f_i f_j = S_e Q f_i Q f_j = S_e f_i Q' Q f_j.$$

If, therefore, we set

$$(27)\qquad f'_j = Q'Qf_j \qquad (j = 1, 2, \cdots, n),$$

we have, on omitting the subscript e in S_e,

$$(28)\qquad Sf_i f'_j = \delta_{ij} \qquad (i, j = 1, 2, \cdots, n).$$

Since $|\,Q'Q\,| \neq 0$, the vectors $f'_1, f'_2, \cdots, f'_n$ form a basis which we say is *reciprocal* to $f_1, f_2, \cdots, f_n$. This definition is of course relative to the fundamental basis since it depends on the function S but, apart from this the basis (f'_i) is uniquely defined when the basis (f_i) is given since the vectors f_i determine P and $Q = P^{-1}$.

The relation between (f'_i) and (f_i) is a reciprocal one; for

$$f'_j = Q'Qf_j = Q'QPe_j = Q'e_j$$

and, if $R = (Q')^{-1}$, we have $f_j = R'Rf'_j$.

If only the set $(f_1, f_2, \cdots, f_r)$ is supposed given originally, and this set of linearly independent vectors is extended by $f_{r+1}, \cdots, f_n$ to form a basis of the n-space, then $f'_{r+1}, \cdots, f'_n$ individually depend on the choice of $f_{r+1}, \cdots, f_n$. But (28) shows that, if $Sf_ix = 0$ $(i = 1, 2, \cdots, r)$, then x belongs to the linear set $(f'_{r+1}, \cdots, f'_n)$; hence this linear set is uniquely determined although the individual members of its basis are not. We may therefore without ambiguity call $\mathfrak{F}' = (f'_{r+1}, \cdots, f'_n)$ reciprocal to $\mathfrak{F} = (f_1, f_2, \cdots, f_r)$; $\mathfrak{F}'$ is then the set of all vectors x for which $Sxy = 0$ whenever y belongs to $\mathfrak{F}$.

In a later chapter we shall require the following lemma.

LEMMA 1. *If* $(f_1, f_2, \cdots, f_r)$ *and* $(f'_{r+1}, \cdots, f'_n)$ *are reciprocal, so also are* $(B^{-1}f_1, B^{-1}f_2, \cdots, B^{-1}f_r)$ *and* $(B'f'_{r+1}, B'f'_{r+2}, \cdots, B'f'_n)$ *where* B *is any non-singular matrix.*

For $SB'f'_iB^{-1}f_j = Sf'_iBB^{-1}f_j = Sf'_if_j = \delta_{ij}$.

Reciprocal bases have a close connection with reciprocal or inverse matrices in terms of which they might have been defined. If P is non-singular and $Pe_i = f_i$ as above, then $P = \Sigma f_iSe_i(\ )$ and, if $Q = \Sigma e_iSf'_i(\ )$, then

$$PQ = \Sigma e_iSf'_if_jSe_j(\ ) = \Sigma\delta_{ij}e_jSe_j(\ ) = 1$$

so that $Q = P^{-1}$.

If $QQ' = 1$, the bases (f_i) and (f'_i) are identical and $Sf_if_j = \delta_{ij}$ for all i and j; the basis is then said to be *orthogonal* as is also the matrix Q. The inverse of an orthogonal matrix and the product of two or more orthogonal matrices are orthogonal; for, if $RR' = 1$,

$$(RQ)(RQ)' = RQQ'R' = RR' = 1.$$

Suppose that $h_1, h_2, \cdots, h_r$ are real vectors which are linearly independent and for which $Sh_ih_j = \delta_{ij}$ $(i \neq j)$; since h_i is real, we have $Sh_ih_i \neq 0$. If $r < n$, we can always find a real vector x which is not in the linear set $(h_1, \cdots, h_r)$ and, if we put

$$h_{r+1} = x - \sum_1^r h_iSh_ix/Sh_ih_i,$$

then $h_{r+1} \neq 0$ and $Sh_ih_{r+1} = 0$ $(i = 1, 2, \cdots, r)$. Hence we can extend the original set to form a basis of the fundamental n-space. If we set $f_i = h_i/(Sh_ih_i)^{\frac{1}{2}}$, then $Sf_if_j = \delta_{ij}$ even when $i = j$; this modified basis is called an *orthogonal basis* of the set.

If the vectors h_i are not necessarily real, it is not evident that x can be chosen so that $Sh_{r+1}h_{r+1} \neq 0$ when $Sh_ih_i \neq 0$ $(i = 1, 2, \cdots, r)$. This may be shown as follows. In the first place we cannot have $Syh_{r+1} = 0$ for every y, and hence $Sh_{r+1}h_{r+1} \neq 0$ when $r = n - 1$. Suppose now that for every choice of x we have $Sh_{r+1}h_{r+1} = 0$; we can then choose a basis $h_{r+1}, \cdots, h_n$ supplementary to $h_1, \cdots, h_r$ such that $Sh_ih_i = 0$ $(i = r + 1, \cdots, n)$ and $Sh_ih_j = 0$ $(i = r + 1, \cdots, n;\ j = 1, 2, \cdots, r)$. Since we cannot have $Sh_{r+1}h_i = 0$ for every h_i of the basis of the n-space, this scalar must be different from 0 for some value of $i > r$, say $r + k$. If we then put $h'_{r+1} = h_{r+1} + h_{r+k}$ in place of h_{r+1}, we have $Sh_ih'_{r+1} = 0$ $(i = 1, 2, \cdots, r)$ as before and also

$$\begin{aligned} Sh'_{r+1}h'_{r+1} &= Sh_{r+1}h_{r+1} + Sh_{r+k}h_{r+k} + 2Sh_{r+1}h_{r+k} \\ &= 2Sh_{r+1}h_{r+k} \neq 0. \end{aligned}$$

We can therefore extend the basis in the manner indicated for real vectors even when the vectors are complex.

When complex coordinates are in question the following lemma is useful; it contains the case discussed above when the vectors used are real.

LEMMA 2. *When a linear set of order r is given, it is always possible to choose a basis $g_1, g_2, \cdots, g_n$ of the fundamental space such that $g_1, \cdots, g_r$ is a basis of the given set and such that $Sg_i\bar{g}_j = \delta_{ij}$ where $\bar{g}_j$ is the vector whose coordinates are the conjugates of the coordinates of g_j when expressed in terms of the fundamental basis.*

The proof is a slight modification of the one already given for the real case. Suppose that $g_1, \cdots, g_s$ are chosen so that $Sg_i\bar{g}_j = \delta_{ij}$ $(i, j = 1, 2, \cdots, s)$ and such that $(g_1, \cdots, g_s)$ lies in the given set when $s < r$ and when $s > r$, then $g_1, \cdots, g_r$ is a basis of this set. We now put

$$g'_{s+1} = x - \sum_1^s g_iSg_ix/S\bar{g}_ig_i$$

which is not 0 provided x is not in $(g_1, \cdots, g_s)$ and, if $s < r$, will lie in the given set provided x does. We may then put

$$g_{s+1} = g'_{s+1}/(Sg'_{s+1}\bar{g}_{s+1})^{\frac{1}{2}}$$

and the lemma follows readily by induction.

If U is the matrix Σe_iSg_i, then $\bar{U} = \Sigma e_iS\bar{g}_i$ and

$$UU' = 1. \tag{29}$$

Such a matrix is called a *unitary* matrix and the basis $g_1, g_2, \cdots, g_n$ is called a unitary basis. A real unitary matrix is of course orthogonal.

1.10 The rank of a matrix. Let $A = \| a_{ij} \|$ be a matrix and set (cf. (12) §1.03)

$$h_i = Ae_i = a_{ji}e_j;$$

then, if

$$x = \Sigma\xi_i e_i = \Sigma e_i Se_i x$$

is any vector, we have

$$Ax = A\Sigma e_i Se_i x = \Sigma Ae_i Se_i x$$

or

$$(30) \qquad Ax = \sum_1^n h_i Se_i x.$$

Any expression of the form $Ax = \sum_1^m a_i Sb_i x$, where a_i, b_i are constant vectors, is a linear homogeneous vector function of x. Here (30) shows that it is never necessary to take $m > n$, but it is sometimes convenient to do so. When we are interested mainly in the matrix and not in x, we may write $A = \Sigma a_i Sb_i(\ )$ or, omitting the brackets, merely

$$(31) \qquad A = \Sigma a_i Sb_i.$$

It follows readily from the definition of the transverse that

$$(32) \qquad A' = \Sigma b_i Sa_i.$$

No matter what vector x is, Ax, being equal to $\Sigma a_i Sb_i x$, is linearly dependent on $a_1, a_2, \cdots, a_m$ or, if the form (30) is used, on $h_1, h_2, \cdots, h_n$. When $|A| \neq 0$, we have seen in Theorem 4 that the h's are linearly independent but, if A is singular, there are linear relations connecting them, and the order of the linear set $(a_1, a_2, \cdots, a_m)$ is less than n.

Suppose in (31) that the a's are not linearly independent, say

$$a_s = \alpha_1 a_1 + \alpha_2 a_2 + \cdots + \alpha_{s-1} a_{s-1},$$

then on substituting this value of a_s in (31) we have

$$A = a_1 S(b_1 + \alpha_1 b_s) + \cdots + a_{s-1} S(b_{s-1} + \alpha_{s-1} b_s) + \sum_{s+1}^m a_i Sb_i,$$

an expression similar to (31) but having at least one term less. A similar reduction can be carried out if the b's are not linearly independent. After a finite number of repetitions of this process we shall finally reach a form

$$(33) \qquad A = \sum_1^r c_i Sd_i$$

in which $c_1, c_2, \cdots, c_r$ are linearly independent and also $d_1, d_2, \cdots, d_r$. The integer r is called the *rank* of A.

It is clear that the value of r is independent of the manner in which the reduction to the form (33) is carried out since it is the order of the linear set $(Ae_1, Ae_2, \cdots, Ae_n)$. We shall, however, give a proof of this which incidently yields some important information regarding the nature of A.

Suppose that by any method we have arrived at two forms of A

$$A = \sum_1^r c_i S d_i = \sum_1^s p_i S q_i,$$

where $(c_1, c_2, \cdots, c_r)$ and $(d_1, d_2, \cdots, d_r)$ are spaces of order r and $(p_1, p_2, \cdots, p_s)$, $(q_1, q_2, \cdots, q_s)$ spaces of order s, and let $(c'_{r+1}, c'_{r+2}, \cdots, c'_n), \cdots,$ $(q'_{s+1}, q'_{s+2}, \cdots, q'_n)$ be the corresponding reciprocal spaces. Then

$$Aq'_j = \sum_1^s p_i S q_i q'_j = p_j \qquad (j = 1, 2, \cdots, s)$$

and also $Aq'_j = \Sigma\, c_i S d_i q'_j$. Hence each p_j lies in $(c_1, c_2, \cdots, c_r)$. Similarly each c_i lies in $(p_1, p_2, \cdots, p_s)$ so that these two subspaces are the same and, in particular, their orders are equal, that is, $r = s$. The same discussion with A' in place of A shows that $(d_1, d_2, \cdots, d_r)$ and $(q_1, q_2, \cdots, q_s)$ are the same. We shall call the spaces $\mathfrak{G}_l = (c_1, c_2, \cdots, c_r)$, $\mathfrak{G}_r = (d_1, d_2, \cdots, d_r)$ the left and right *grounds* of A, and the total space $\mathfrak{G} = (c_1, \cdots, c_r, d_1, \cdots, d_r)$ will be called the (total) ground of A.

If x is any vector in the subspace $\mathfrak{N}_r = (d'_{r+1}, d'_{r+2}, \cdots, d'_n)$ reciprocal to $\mathfrak{G}_r$, then $Ax = 0$ since $Sd_i d'_j = 0$ $(i \neq j)$. Conversely, if

$$0 = Ax = \Sigma\, c_i S d_i x,$$

each multiplier $Sd_i x$ must be 0 since the c's are linearly independent; hence every solution of $Ax = 0$ lies in $\mathfrak{N}_r$. Similarly every solution of $A'x = 0$ lies in $\mathfrak{N}_l = (c'_{r+1}, c'_{r+2}, \cdots, c'_n)$. We call $\mathfrak{N}_r$ and $\mathfrak{N}_l$ the right and left *nullspaces* of A; their order, $n - r$, is called the *nullity* of A.

We may summarize these results as follows.

THEOREM 7. *If a matrix A is expressed in the form $\sum_1^r a_i S b_i$, where $\mathfrak{G}_l = (a_1, a_2, \cdots, a_r)$ and $\mathfrak{G}_r = (b_1, b_2, \cdots, b_r)$ define spaces of order r, then, no matter how the reduction to this form is carried out, the spaces $\mathfrak{G}_l$ and $\mathfrak{G}_r$ are always the same. Further, if $\mathfrak{N}_l$ and $\mathfrak{N}_r$ are the spaces of order $n - r$ reciprocal to $\mathfrak{G}_l$ and $\mathfrak{G}_r$, respectively, every solution of $Ax = 0$ lies in $\mathfrak{N}_r$ and every solution of $A'x = 0$ in $\mathfrak{N}_l$.*

The following theorem is readily deduced from Theorem 7 and its proof is left to the reader.

THEOREM 8. *If A, B are matrices of rank r, s, the rank of $A + B$ is not greater than $r + s$ and the rank of AB is not greater than the smaller of r and s.*

1.11 Linear dependence. The definition of the rank of a matrix in the preceding section was made in terms of the linear dependence of vectors associated with the matrix. In this section we consider briefly the theory of linear dependence introducing incidentally a notation which we shall require later.

Let $x_i = \sum_{j=1}^{n} \xi_{ij} e_j$ $(i = 1, 2, \cdots, r; r \leq n)$ be a set of r vectors. From the rectangular array of their coordinates

$$\begin{array}{cccc} \xi_{11} & \xi_{12} & \cdots & \xi_{1n} \\ \xi_{21} & \xi_{22} & \cdots & \xi_{2n} \\ \cdot & \cdot & \cdots & \cdot \\ \xi_{r1} & \xi_{r2} & \cdots & \xi_{rn} \end{array} \tag{34}$$

there can be formed $n!/r!(n - r)!$ different determinants of order r by choosing r columns out of (34), these columns being taken in their natural order. If these determinants are arranged in some definite order, we may regard them as the coordinates of a vector in space of order $n!/r!(n - r)!$ and, when this is done, we shall denote this vector by[4]

$$| x_1 x_2 \cdots x_r | \tag{35}$$

and call it a *pure vector* of *grade* r. It follows from this definition that $| x_1 x_2 \cdots x_r |$ has many of the properties of a determinant; its sign is changed if two x's are interchanged, it vanishes when two x's are equal and, if λ and μ are scalars,

$$| (\lambda x_1 + \mu x_1') \, x_2 \cdots x_r | = \lambda \, | x_1 x_2 \cdots x_r | + \mu \, | x_1' x_2 \cdots x_r |. \tag{36}$$

If we replace the x's in (35) by r different units $e_{i_1}, e_{i_2}, \cdots, e_{i_r}$, the result is clearly not 0: we thus obtain $\binom{n}{r}$ vectors which we shall call the fundamental unit vectors of grade r; and any linear combination of these units, say

$$\Sigma \, \xi_{i_1 i_2 \ldots i_r} | e_{i_1} e_{i_2} \cdots e_{i_r} |,$$

is called a vector of grade r. It should be noticed that not every vector is a pure vector except when r equals 1 or n.

If we replace x_i by $\Sigma \, \xi_{ij} e_j$ in (35), we get

$$| x_1 x_2 \cdots x_r | = \Sigma \, \xi_{1j_1} \xi_{2j_r} \cdots \xi_{rj_r} | e_{j_1} e_{j_2} \cdots e_{j_r} |,$$

where the summation extends over all permutations $j_1, j_2, \cdots, j_r$ of $1, 2, \cdots, n$ taken r at a time. This summation may be effected by grouping together the

[4] If it had been advisable to use here the indeterminate product of Grassmann, (35) would appear as a determinant in much the ordinary sense (cf. §5.09).

sets $j_1, j_2, \cdots, j_r$ which are permutations of the same combination $i_1, i_2, \cdots, i_r$, whose members may be taken to be arranged in natural order, and then summing these partial sums over all possible combinations $i_1, i_2, \cdots, i_r$. Taking the first step only we have

$$\Sigma\, \xi_{1j_1}\xi_{2j_2} \cdots \xi_{rj_r} \,|\, e_{j_1}e_{j_2} \cdots e_{j_r} \,| = \Sigma\, \delta^{i_1 \cdots i_r}_{j_1 \cdots j_r}\, \xi_{1j_1} \cdots \xi_{rj_r} \,|\, e_{i_1}e_{i_2} \cdots e_{i_r} \,|$$

where $\delta^{i_1 \cdots i_r}_{j_1 \cdots j_r}$ is the sign corresponding to the permutations $\binom{i_1 i_2 \cdots i_r}{j_1 j_2 \cdots j_r}$ and this equals $|\, \xi_{1i_1} \cdots \xi_{ri_r} \,|\,|\, e_{i_1} \cdots e_{i_r} \,|$. We have therefore

$$(37) \qquad |\, x_1x_2 \cdots x_r \,| = \sum_{(i)}^{*} |\, \xi_{1i_1}\xi_{2i_2} \cdots \xi_{ri_r} \,|\,|\, e_{i_1}e_{i_2} \cdots e_{i_r} \,|,$$

where the asterisk on Σ indicates that the sum is taken over all r-combinations of $1, 2, \cdots, n$ each combination being arranged in natural order.

THEOREM 9 $|\, x_1x_2 \cdots x_r \,| = 0$ *if, and only if,* $x_1, x_2, \cdots, x_r$ *are linearly dependent.*

The first part of this theorem is an immediate consequence of (36). To prove the converse it is sufficient to show that, if $|\, x_1x_2 \cdots x_{r-1} \,| \neq 0$, then there exist scalars $\alpha_1, \alpha_2, \cdots, \alpha_{r-1}$ such that

$$x_r = \alpha_1x_1 + \alpha_2x_2 + \cdots + \alpha_{r-1}x_{r-1}.$$

Let $x_i = \sum_j \xi_{ij}e_j$. Since $|\, x_1x_2 \cdots x_{r-1} \,| \neq 0$, at least one of its coordinates is not 0, and for convenience we may suppose without loss of generality that

$$(38) \qquad |\, \xi_{11}\xi_{22} \cdots \xi_{r-1,\, r-1} \,| \neq 0.$$

Since $|\, x_1x_2 \cdots x_r \,| = 0$, all its coordinates equal 0 and in particular

$$|\, \xi_{11}\xi_{22} \cdots \xi_{r-1,\, r-1}\xi_{ri} \,| = 0 \qquad (i = 1, 2, \cdots, n).$$

If we expand this determinant according to the elements of its last column, we get a relation of the form

$$\beta_1\xi_{ri} + \beta_2\xi_{1i} + \cdots + \beta_r\xi_{r-1,\, i} = 0$$

where the β's are independent of i and $\beta_1 \neq 0$ by (38). Hence we may write

$$(39) \qquad \xi_{ri} = \alpha_1\xi_{1i} + \cdots + \alpha_{r-1}\xi_{r-1,\, i} \qquad (i = 1, 2, \cdots, n)$$

the α's being independent of i. Multiplying (39) by e_i and summing with regard to i, we have

$$x_r = \alpha_1x_1 + \cdots + \alpha_{r-1}x_{r-1},$$

which proves the theorem.

If $(a_1, a_2, \cdots, a_m)$ is a linear set of order r, then some set of r a's form a basis, that is, are linearly independent while each of the other a's is linearly dependent on them. By a change of notation, if necessary, we may take $a_1, a_2, \cdots, a_r$ as this basis and write

$$a_{r+i} = \sum_{j=1}^{r} \beta_{ij} a_j, \qquad (i = 1, 2, \cdots, m - r). \tag{40}$$

We shall now discuss the general form of all linear relations among the a's in terms of the special relations (40); and in doing so we may assume the order of the space to be equal to or greater than m since we may consider any given space as a subspace of one of arbitrarily higher dimensionality.

Let

$$\sum_{1}^{m} \gamma_j a_j = 0 \tag{41}$$

be a relation connecting the a's and set

$$c = \sum_{1}^{m} \gamma_j e_j.$$

Then (40), considered as a special case of (41), corresponds to setting for c

$$c_i = -\sum_{j=1}^{r} \beta_{ij} e_j + e_{r+i}, \qquad (i = 1, 2, \cdots, m - r); \tag{42}$$

and there is clearly no linear relation connecting these vectors so that they define a linear set of order $m - r$. Using (40) in (41) we have

$$\sum_{j=1}^{r} \left(\gamma_j + \sum_{i=1}^{m-r} \gamma_{r+i} \beta_{ij} \right) a_j = 0$$

and, since $a_1, a_2, \cdots, a_r$ are linearly independent, we have

$$j = -\sum_{i=1}^{m-r} \beta_{ij} \gamma_{r+i} \qquad (j = 1, 2, \cdots, r)$$

whence

$$c = \sum_{1}^{m} \gamma_j e_j = -\sum_{i=1}^{m-r} \gamma_{r+i} \sum_{j=1}^{r} \beta_{ij} e_j + \sum_{i=1}^{m-r} \gamma_{r+i} e_{r+i} = \sum_{i=1}^{m-r} \gamma_{r+i} c_i, \tag{43}$$

so that c is linearly dependent on $c_1, c_2, \cdots, c_{m-r}$. Conversely, on retracing these steps in the reverse order we see that, if c is linearly dependent on these vectors, so that γ_{r+i} $(i = 1, 2, \cdots, m - r)$ are known, then from (43) the γ_j $(j = 1, 2, \cdots, r)$ are defined in such a way that $c = \sum_{1}^{m} \gamma_j e_j$ and $\sum_{1}^{m} \gamma_j a_j = 0$. We have therefore the following theorem.

THEOREM 10. *If* $a_1, a_2, \cdots, a_m$ *is a linear set of order* r, *there exist* $m - r$ *linear relations* $\sum_{j=1}^{m} \gamma_{ij} a_j = 0$ $(i = 1, 2, \cdots, m - r)$ *such that* (i) *the vectors* $c_i = \sum_{j=1}^{m} \gamma_{ij} e_j$ *are linearly independent and* (ii) *if* $\Sigma\, \gamma_j a_j = 0$ *is any linear relation connecting the* a's, *and if* $c = \Sigma\, \gamma_j e_j$, *then* c *belongs to the linear set* $(c_1, c_2, \cdots, c_{m-r})$.

This result can be translated immediately in terms concerning the solution of a system of ordinary linear equations or in terms of matrices. If $a_j = \sum_i a_{ji} e_i$, then (41) may be written

$$
\begin{array}{c}
a_{11}\gamma_1 + a_{21}\gamma_2 + \cdots + a_{m1}\gamma_m = 0 \\
\cdots\cdots\cdots\cdots\cdots\cdots\cdots\cdots \\
\cdots\cdots\cdots\cdots\cdots\cdots\cdots\cdots \\
a_{1n}\gamma_1 + a_{2n}\gamma_2 + \cdots + a_{mn}\gamma_m = 0
\end{array}
\tag{44}
$$

a system of linear homogeneous equations in the unknowns $\gamma_1, \gamma_2, \cdots, \gamma_m$. Hence (44) has solutions for which some $\gamma_i \neq 0$ if, and only if, the rank r of the array

$$
\begin{array}{cccc}
a_{11} & a_{21} & \cdots & a_{m1} \\
a_{12} & a_{22} & \cdots & a_{m2} \\
\cdot & \cdot & \cdots & \cdot \\
a_{1n} & a_{2n} & \cdots & a_{mn}
\end{array}
\tag{45}
$$

is less then m and, when this condition is satisfied, every solution is linearly dependent on the set of $m - r$ solutions given by (42) which are found by the method given in the discussion of Theorem 9.

Again, if we make (45) a square array by the introduction of columns or rows of zeros and set $A = \| a_{ij} \|$, $c = \Sigma\, \gamma_i e_i$, then (41) becomes $A'c = 0$ and Theorem 10 may therefore be interpreted as giving the properties of the null-space of A' which were derived in §1.10.

CHAPTER II

ALGEBRAIC OPERATIONS WITH MATRICES. THE CHARACTERISTIC EQUATION

2.01 Identities. The following elementary considerations enable us to carry over a number of results of ordinary scalar algebra into the algebra of matrices. Suppose $f(\lambda_1, \lambda_2, \cdots, \lambda_r)$, $g(\lambda_1, \lambda_2, \cdots, \lambda_r)$ are integral algebraic functions of the scalar variables λ_i with scalar coefficients, and suppose that

$$f(\lambda_1, \lambda_2, \cdots, \lambda_r) = g(\lambda_1, \lambda_2, \cdots, \lambda_r)$$

is an algebraic identity; then, when $f(\lambda_1, \cdots, \lambda_r) - g(\lambda_1, \cdots, \lambda_r)$ is reduced to the standard form of a polynomial, the coefficients of the various powers of the λ's are zero. In carrying out this reduction no properties of the λ's are used other than those which state that they obey the laws of scalar multiplication and addition: if then we replace $\lambda_1, \lambda_2, \cdots, \lambda_r$ by *commutative* matrices $x_1, x_2, \cdots, x_r$, the reduction to the form 0 is still valid step by step and hence

$$f(x_1, x_2, \cdots, x_r) = g(x_1, x_2, \cdots, x_r).$$

An elementary example of this is

$$(1 - x^2) = (1 - x)(1 + x)$$

or, when $xy = yx$,

$$x^2 - y^2 = (x - y)(x + y).$$

Here, if $xy \neq yx$, the reader should notice that the analogue of the algebraic identity becomes

$$x^2 - y^2 = x(x + y) - (x + y)y,$$

which may also be written $x^2 - y^2 = (x - y)(x + y) + (yx - xy)$.

2.02 Matric polynomials in a scalar variable. By a matric polynomial in a scalar variable λ is meant a matrix that can be expressed in the form

$$(1) \qquad P(\lambda) = p_0\lambda^r + p_1\lambda^{r+1} + \cdots + p_r \qquad (p_0 \neq 0),$$

where $p_0, p_1, \ldots, p_r$ are constant matrices. The coordinates of $P(\lambda)$ are scalar polynomials in λ and hence, if

$$(2) \qquad Q(\lambda) = q_0\lambda^s + q_1\lambda^{s-1} + \cdots + q_s \qquad (q_0 \neq 0)$$

is also a matric polynomial, $P(\lambda) \equiv Q(\lambda)$ if, and only if, $r = s$ and the coefficients of corresponding powers of λ are equal, that is, $p_i = q_i$ $(i = 1, 2, \ldots, r)$. If $|q_0| \neq 0$, the degree of the product $P(\lambda)Q(\lambda)$ (or $Q(\lambda)P(\lambda)$) is exactly $r + s$ since the coefficient of the highest power λ^{r+s} which occurs in the product is p_0q_0

(or q_0p_0) which cannot be 0 if $p_0 \neq 0$ and $|q_0| \neq 0$. If, however, both $|p_0|$ and $|q_0|$ are 0, the degree of the product may well be less than $r + s$, as is seen from the examples

$$(e_{11}\lambda + 1)\,(e_{22}\lambda + 1) = e_{11}e_{22}\lambda^2 + (e_{11} + e_{22})\lambda + 1 = (e_{11} + e_{22})\lambda + 1,$$

$$\begin{Vmatrix} \lambda & 1 \\ \lambda & 1 \end{Vmatrix} \begin{Vmatrix} 1 & -1 \\ -\lambda & \lambda \end{Vmatrix} = 0.$$

Another noteworthy difference between matric and scalar polynomials is that, when the determinant of a matric polynomial is a constant different from 0, its inverse is also a matric polynomial: for instance

$$(e_{12}\lambda + 1)^{-1} = -e_{12}\lambda + 1,$$
$$[(e_{12} + e_{23})\lambda + 1]^{-1} = e_{13}\lambda^2 - (e_{12} + e_{23})\lambda + 1.$$

We shall call such polynomials *elementary polynomials*.

2.03 **The division transformation.** The greater part of the theory of the division transformation can be extended from ordinary algebra to the algebra of matrices; the main precaution that must be taken is that it must not be assumed that every element of the algebra has an inverse and that due allowance must be made for the peculiarities introduced by the lack of commutativity in multiplication.

THEOREM 1. *If $P(\lambda)$ and $Q(\lambda)$ are the polynomials defined by* (1) *and* (2), *and if $|q_0| \neq 0$, there exist unique polynomials $S(\lambda)$, $R(\lambda)$, $S_1(\lambda)$, $R_1(\lambda)$, of which S and S_1 if not zero, are of degree $r - s$ and the degrees of R and R_1 are $s - 1$ at most, such that*

$$P(\lambda) \equiv S(\lambda)Q(\lambda) + R(\lambda) \equiv Q(\lambda)S_1(\lambda) + R_1(\lambda).$$

If $r < s$, we may take $S_1 = S = 0$ and $R_1 = R = P$; in so far as the existence of these polynomials is concerned the theorem is therefore true in this case. We shall now assume as a basis for a proof by induction that the theorem is true for polynomials of degree less than r and that $r \leq s$. Since $|q_0| \neq 0$, q_0^{-1} exists and, as in ordinary scalar division, we have

$$P(\lambda) - p_0q_0^{-1}\lambda^{r-s}Q(\lambda) = (p_1 - p_0q_0^{-1}q_1)\lambda^{r-1} + \cdots = P_1(\lambda).$$

Since the degree of P_1 is less than r, we have by hypothesis $P_1(\lambda) = P_2(\lambda)Q(\lambda) + R(\lambda)$, the degrees of P_2 and R being less, respectively, than $r - s$ and s; hence

$$P(\lambda) = (p_0q_0^{-1}\lambda^{r-s} + P_2(\lambda))Q(\lambda) + R(\lambda) = S(\lambda)Q(\lambda) + R(\lambda)$$

as required by the theorem. The existence of the right hand quotient and remainder follows in the same way.

It remains to prove the uniqueness of S and R. Suppose, if possible, that $P = SQ + R = TQ + U$ where R and S are as above and T, U are polynomials the degree of U being less than s; then $(S - T)Q = U - R$. If $S - T \neq 0$, then, since $|q_0| \neq 0$, the degree of the polynomial $(S - T)Q$ is at least as great as that of Q and is therefore greater than the degree of $U - R$. It follows immediately that $S - T = 0$, and hence also $U - R = 0$; which completes the proof of the theorem.

If Q is a scalar polynomial, that is, if its coefficients q are scalars, then $S = S_1$, $R = R_1$; and, if the division is exact, then $Q(\lambda)$ is a factor of each of the coordinates of $P(\lambda)$.

THEOREM 2. *If the matric polynomial* (1) *is divided on the right by* $\lambda - a$, *the remainder is*

$$p_0a^r + p_1a^{r-1} + \cdots + p_r$$

and, if it is divided on the left, the remainder is

$$a^rp_0 + a^{r-1}p_1 + \cdots + p_r.$$

As in ordinary algebra the proof follows immediately from the identity

$$\lambda^s - a^s = (\lambda - a)(\lambda^{s-1} + \lambda^{s-2}a + \cdots + a^{s-1})$$

in which the order of the factors is immaterial since λ is a scalar.

If $P(\lambda)$ is a scalar polynomial, the right and left remainders are the same and are conveniently denoted by $P(a)$.

2.04 Theorem 1 of the preceding section holds true as regards the existence of S, S_1, R, R_1, and the degree of R, R_1 even when $|q_0| = 0$ provided $|Q(\lambda)| \neq 0$. Suppose the rank of q_0 is $t < n$; then by §1.10 it has the form $\sum_1^t \alpha_i S\beta_i$ or, say, $h\left(\sum_1^t e_{ii}\right)k$ where h and k are non-singular matrices for which $he_i = \alpha_i$, $k'e_i = \beta_i$ $(i = 1, 2, \cdots, t)$. If $c_1 = \sum_{t+1}^n e_{ii}$, then

$$Q_1 = (c_1\lambda + 1)h^{-1}Q \tag{3}$$

is a polynomial whose degree is not higher than the degree s of Q since $c_1h^{-1}q_0 = 0$ so that the term in λ^{s+1} is absent. Now, if $\eta = |h^{-1}|$, then

$$|Q_1| = |c_1\lambda + 1|\,|h^{-1}|\,|Q| = (1 + \lambda)^{n-t}\eta\,|Q|,$$

so that the degree of $|Q_1|$ is greater than that of $|Q|$ by $n - t$. If the leading coefficient of Q_1 is singular, this process may be repeated, and so on, giving $Q_1, Q_2, \cdots$, where the degree of $|Q_i|$ is greater than that of $|Q_{i-1}|$. But the degree of each Q_i is less than or equal to s and the degree of the determinant of a polynomial of the sth degree cannot exceed ns. Hence at some stage the leading coefficient of, say, Q_j is not singular and, from the law of formation (3) of the successive Q's, we have $Q_j(\lambda) = H(\lambda)Q(\lambda)$, where $H(\lambda)$ is a matric polynomial.

By Theorem 1, Q_j taking the place of Q, we can find S^* and R, the latter of degree $s - 1$ at most, such that

$$P(\lambda) = S^*(\lambda)H(\lambda)Q(\lambda) + R(\lambda) = S(\lambda)Q(\lambda) + R(\lambda).$$

The theorem is therefore true even if $|q_0| = 0$ except that the quotient and remainder are not necessarily unique and the degree of S may be greater than $r - s$, as is shown by taking $P = \lambda^2 - 1$, $Q = e_{11}\lambda + 1$, when we have

$$P = (e_{22}\lambda^2 + e_{11}\lambda - 1)Q = (e_{22}\lambda^2 + e_{11}\lambda - 1 + e_{12})Q - e_{12}.$$

2.05 **The characteristic equation.** If x is a matrix, the scalar polynomial

(4) $$f(\lambda) = |\lambda - x| = \lambda^n + a_1\lambda^{n-1} + \cdots + a_n$$

is called the *characteristic function* corresponding to x. We have already seen (§1.05 (15)) that the product of a matrix and its adjoint equals its determinant; hence

$$(\lambda - x) \text{ adj } (\lambda - x) = |\lambda - x| = f(\lambda).$$

It follows that the polynomial $f(\lambda)$ is exactly divisible by $\lambda - x$ so that by the remainder theorem (§2.03, Theorem 2)

(5) $$f(x) = 0.$$

As a simple example of this we may take $x = \begin{Vmatrix} \alpha & \beta \\ \gamma & \delta \end{Vmatrix}$. Here

$$f(\lambda) = (\lambda - \alpha)(\lambda - \delta) - \beta\gamma = \lambda^2 - (\alpha + \delta)\lambda + \alpha\delta - \beta\gamma,$$

and

$$f(x) = \begin{Vmatrix} \alpha^2 + \beta\gamma & \alpha\beta + \beta\delta \\ \gamma\alpha + \delta\gamma & \gamma\beta + \delta^2 \end{Vmatrix} - (\alpha + \delta) \begin{Vmatrix} \alpha & \beta \\ \gamma & \delta \end{Vmatrix} + (\alpha\delta - \beta\gamma) \begin{Vmatrix} 1 & 0 \\ 0 & 1 \end{Vmatrix} = 0.$$

The following theorem is an important extension of this result.

Theorem 3. *If $f(\lambda) = |\lambda - x|$ and $\theta(\lambda)$ is the highest common factor of the first minors of $|\lambda - x|$, and if*

(6) $$\varphi(\lambda) = f(\lambda)/\theta(\lambda),$$

the leading coefficient of $\theta(\lambda)$ being 1 (*and therefore also that of $\varphi(\lambda)$*), *then*

(i) $\varphi(x) = 0$;

(ii) *if $\psi(\lambda)$ is any scalar polynomial such that $\psi(x) = 0$, then $\varphi(\lambda)$ is a factor of $\psi(\lambda)$, that is, $\varphi(\lambda)$ is the scalar polynomial of lowest degree and with leading coefficient* 1 *such that $\varphi(x) = 0$;*

(iii) *every root of $f(\lambda)$ is a root of $\varphi(\lambda)$.*

The coordinates of adj$(\lambda - x)$ are the first minors of $|\lambda - x|$ and therefore by hypothesis $[\text{adj}(\lambda - x)]/\theta(\lambda)$ is integral; also

$$\frac{\text{adj}(\lambda - x)}{\theta(\lambda)}(\lambda - x) = \frac{f(\lambda)}{\theta(\lambda)} = \varphi(\lambda);$$

hence $\varphi(x) = 0$ by the remainder theorem.

If $\psi(\lambda)$ is any scalar polynomial for which $\psi(x) = 0$, we can find scalar polynomials $M(\lambda)$, $N(\lambda)$ such that $M(\lambda)\varphi(\lambda) + N(\lambda)\psi(\lambda) \equiv \zeta(\lambda)$, where $\zeta(\lambda)$ is the highest common factor of φ and ψ. Substituting x for λ in this scalar identity and using $\varphi(x) = 0 = \psi(x)$ we have $\zeta(x) = 0$; if, therefore, $\psi(x) = 0$ is a scalar equation of lowest degree satisfied by x, we must have $\psi(\lambda) = \zeta(\lambda)$, apart from a constant factor, so that $\psi(\lambda)$ is a factor of $\varphi(\lambda)$, say

$$\varphi(\lambda) = h(\lambda)\psi(\lambda). \tag{7}$$

Since $\psi(x) = 0$, $\lambda - x$ is a factor of $\psi(\lambda)$, say $\psi(\lambda) = (\lambda - x)g(\lambda)$, where g is a matric polynomial; hence

$$\psi(\lambda) = \frac{\varphi(\lambda)}{h(\lambda)} = \frac{f(\lambda)}{h(\lambda)\theta(\lambda)} = (\lambda - x)g(\lambda).$$

Hence

$$g(\lambda) = \frac{f(\lambda)}{\theta(\lambda)h(\lambda)(\lambda - x)} = \frac{\operatorname{adj}(\lambda - x)}{\theta(\lambda)h(\lambda)}$$

and this cannot be integral unless $h(\lambda)$ is a constant in view of the fact that $\theta(\lambda)$ is the highest common factor of the coordinates of $\operatorname{adj}(\lambda - x)$; it follows that $\psi(\lambda)$ differs from $\varphi(\lambda)$ by at most a constant factor.

A repetition of the first part of this argument shows that, if $\psi(x) = 0$ is any scalar equation satisfied by x, then $\varphi(\lambda)$ is a factor of $\psi(\lambda)$.

It remains to show that every root of $f(\lambda)$ is a root of $\varphi(\lambda)$. If λ_1 is any root of $f(\lambda) = |\lambda - x|$, then from $\varphi(\lambda) = g(\lambda)(\lambda - x)$ we have

$$\varphi(\lambda_1) = g(\lambda_1)(\lambda_1 - x)$$

so that the determinant, $[\varphi(\lambda_1)]^n$, of the scalar matrix $\varphi(\lambda_1)$ equals $|g(\lambda_1)|\,|\lambda_1 - x|$, which vanishes since $|\lambda_1 - x| = f(\lambda_1)$. This is only possible if $\varphi(\lambda_1) = 0$, that is, if every root of $f(\lambda)$ is also a root of $\varphi(\lambda)$.

The roots of $f(\lambda)$ are also called the *roots*[1] of x, $\varphi(\lambda)$ is called the *reduced characteristic function* of x, and $\varphi(x) = 0$ the *reduced equation* of x.

2.06 A few simple results are conveniently given at this point although they are for the most part merely particular cases of later theorems. If $g(\lambda)$ is a scalar polynomial, then on dividing by $\varphi(\lambda)$, whose degree we shall denote by ν, we may set $g(\lambda) = q(\lambda)\varphi(\lambda) + r(\lambda)$, where q and r are polynomials the degree of r being less than ν. Replacing λ by x in this identity and remembering that $\varphi(x) = 0$, we have[2] $g(x) = r(x)$, that is, any polynomial can be replaced by an equivalent polynomial of degree less than ν.

[1] They are also called the *latent* roots of x.

[2] If $g(\lambda)$ is a matric polynomial whose coefficients are not all commutative with x, the meaning of $g(x)$ is ambiguous; for instance, x may be placed on the right of the coefficients, or it may be put on the left. For such a polynomial we can say in general that it can be replaced by an equal polynomial in which no power of x higher than the $(\nu - 1)$th occurs.

If $g(\lambda)$ is a scalar polynomial which is a factor of $\varphi(\lambda)$, say $\varphi(\lambda) = h(\lambda)g(\lambda)$, then $0 = \varphi(x) = h(x)g(x)$. It follows that $|g(x)| = 0$; for if this were not so, we should have $h(x) = [g(x)]^{-1}\varphi(x) = 0$, whereas x can satisfy no scalar equation of lower degree than φ. Hence, if $g(\lambda)$ is a scalar polynomial which has a factor in common with $\varphi(x)$, then $g(x)$ is singular.

If a scalar polynomial $g(\lambda)$ has no factor in common with $\varphi(\lambda)$, there exist scalar polynomials $M(\lambda)$, $N(\lambda)$ such that $M(\lambda)g(\lambda) + N(\lambda)\varphi(\lambda) \equiv 1$. Hence $M(x)g(x) = 1$, or $[g(x)]^{-1} = M(x)$. It follows immediately that any finite rational function of x with scalar coefficients can be expressed as a scalar polynomial in x of degree $\nu - 1$ at most. It should be noticed carefully however that, if x is a variable matrix, the coefficients of the reduced polynomial will in general contain the variable coordinates of x and will not be integral in these unless the original function is integral. It follows also that $g(x)$ is singular only when $g(\lambda)$ has a factor in common with $\varphi(\lambda)$.

Finally we may notice here that similar matrices have the same reduced equation; for, if g is a scalar polynomial, $g(y^{-1}xy) = y^{-1}g(x)y$. As a particular case of this we have that xy and yx have the same reduced equation if, say, y is non-singular; for $xy = y^{-1}\cdot yx\cdot y$. If both x and y are singular, it can be shown[3] that xy and yx have the same characteristic equation, but not necessarily the same reduced equation as is seen from the example $x = e_{12}$, $y = e_{22}$.

2.07 **Matrices with distinct roots.** Because of its importance and comparative simplicity we shall investigate the form of a matrix all of whose roots are different before considering the general case. Let

$$f(\lambda) = |\lambda - x| = (\lambda - \lambda_1)(\lambda - \lambda_2) \cdots (\lambda - \lambda_n) \tag{8}$$

where no two roots are equal and set

$$f_i(\lambda) = \frac{(\lambda - \lambda_1) \cdots (\lambda - \lambda_{i-1})(\lambda - \lambda_{i+1}) \cdots (\lambda - \lambda_n)}{(\lambda_i - \lambda_1) \cdots (\lambda_i - \lambda_{i-1})(\lambda_i - \lambda_{i+1}) \cdots (\lambda_i - \lambda_n)} = \frac{f(\lambda)/f'(\lambda_i)}{\lambda - \lambda_i}. \tag{9}$$

By the Lagrange interpolation formula $\sum_i f_i(\lambda) \equiv 1$; hence

$$f_1(x) + f_2(x) + \cdots + f_n(x) = 1. \tag{10}$$

Further, $f(\lambda)$ is a factor of $f_i(\lambda)f_j(\lambda)$ $(i \neq j)$ so that

$$f_i(x)f_j(x) = 0 \qquad (i \neq j); \tag{11}$$

hence multiplying (10) by $f_i(x)$ and using (11) we have

$$[f_i(x)]^2 = f_i(x). \tag{12}$$

Again, $(\lambda - \lambda_i)f_i(\lambda) = f(\lambda)/f'(\lambda_i)$; hence $(x - \lambda_i)f_i(x) = 0$, that is,

$$xf_i(x) = \lambda_i f_i(x), \tag{13}$$

[3] For example, by replacing y by $y + \delta$, δ being a scalar, and considering the limiting case when δ approaches 0.

whence, summing with regard to i and using (10), we have

$$x = \lambda_1 f_1(x) + \lambda_2 f_2(x) + \cdots + \lambda_n f_n(x). \tag{14}$$

If we form x^r from (14), r being a positive integer, it is immediately seen from (11) and (12), or from the Lagrange interpolation formula, that

$$x^r = \lambda_1^r f_1 + \lambda_2^r f_2 + \cdots + \lambda_n^r f_n, \tag{15}$$

where f_i stands for $f_i(x)$, and it is easily verified by actual multiplication that, if no root is 0,

$$x^{-1} = \lambda_1^{-1} f_1 + \lambda_2^{-1} f_2 + \cdots + \lambda_n^{-1} f_n$$

so that (15) holds for negative powers also. The matrices f_i are linearly independent. For if $\Sigma \gamma_i f_i = 0$, then

$$0 = f_j \Sigma \gamma_i f_i = \gamma_j f_j^2 = \gamma_j f_j$$

whence every $\gamma_j = 0$ seeing that in the case we are considering $f(\lambda)$ is itself the reduced characteristic function so that $f_j(x) \neq 0$.

From these results we have that, if $g(\lambda)$ is any scalar rational function whose denominator has no factor in common with $\varphi(\lambda)$, then

$$g(x) = g(\lambda_1) f_1 + g(\lambda_2) f_2 + \cdots + g(\lambda_n) f_n. \tag{16}$$

It follows from this that the roots of $g(x)$ are $g(\lambda_i)$ $(i = 1, 2, \cdots, n)$. For setting $y = g(x)$, $\mu_i = g(\lambda_i)$, we have as above

$$\psi(y) = \Sigma \psi(\mu_i) f_i,$$

$\psi(\lambda)$ being a scalar polynomial. Now $\psi(y) f_i = \psi(\mu_i) f_i$; hence, if $\psi(y) = 0$, then also $\psi(\mu_i) = 0$ $(i = 1, 2, \cdots, n)$; and conversely. Hence if the notation is so chosen that $\mu_1, \mu_2, \cdots, \mu_r$ are the distinct values of μ_i, the reduced characteristic function of $y = g(x)$ is $\prod_1^r (\lambda - \mu_i)$.

2.08 If the determinant $|\lambda - x| = f(\lambda)$ is expanded in powers of λ, it is easily seen[4] that the coefficient a_r of λ^{n-r} is $(-1)^r$ times the sum of the principal minors of x of order r; this coefficient is therefore a homogeneous polynomial of degree r in the coordinates of x. In particular, $-a_1$ is the sum of the coordinates in the main diagonal: this sum is called the *trace* of x and is denoted by tr x.

If y is an arbitrary matrix, μ a scalar variable, and $z = x + \mu y$, the coefficients of the characteristic equation of z, say

$$z^n + b_1 z^{n-1} + \cdots + b_n = 0, \tag{17}$$

[4] For instance, by differentiating $|\lambda - x|$ $n - r$ times with respect to λ and then setting $\lambda = 0$.

are polynomials in μ of the form

(18) $$b_s = a_{s0} + \mu a_{s1} + \cdots + \mu^s a_{ss}, \quad (a_{s0} = a_s, a_{00} = 1)$$

and the powers of z are also polynomials in μ, say

(19) $$z^r = x^r + \mu \begin{Bmatrix} x & y \\ r-1 & 1 \end{Bmatrix} + \mu^2 \begin{Bmatrix} x & y \\ r-2 & 2 \end{Bmatrix} + \cdots + \mu^r y^r$$

where $\begin{Bmatrix} x & y \\ s & t \end{Bmatrix}$ is obtained by multiplying s x's and t y's together in every possible way and adding the terms so obtained, e.g.,

$$\begin{Bmatrix} x & y \\ 2 & 1 \end{Bmatrix} = x^2y + xyx + yx^2.$$

If we substitute (18) and (19) in (17) and arrange according to powers of μ, then, since μ is an independent variable, the coefficients of its several powers must be zero. This leads to a series of relations connecting x and y of the form

(20) $$\sum_{i,j} a_{ij} \begin{Bmatrix} x & y \\ n-s-i+j & s-j \end{Bmatrix} = 0 \qquad (s = 0, 1, 2, \cdots)$$

where a_{ij} are the coefficients defined in (18) and $\begin{Bmatrix} x & y \\ n-s-i+j & s-j \end{Bmatrix}$ is replaced by 0 when $j > s$. In particular, if $s = 1$,

$$\begin{Bmatrix} x & y \\ n-1 & 1 \end{Bmatrix} + a_1 \begin{Bmatrix} x & y \\ n-2 & 1 \end{Bmatrix} + \cdots + a_{n-1}y + a_{11}x^{n-1} + \cdots + a_{n1} = 0$$

which, when $xy = yx$, becomes

$$f'(x)y = -(a_{11}x^{n-1} + \cdots + a_{n1}) = g(x).$$

When x has no repeated roots, $f'(\lambda)$ has no root in common with $f(\lambda)$ and $f'(x)$ has an inverse (cf. §2.06) so that $y = g(x)/f'(x)$ which can be expressed as a scalar polynomial in x; and conversely every such polynomial is commutative with x. We therefore have the following theorem:

THEOREM 4. *If x has no multiple roots, the only matrices commutative with it are scalar polynomials in x.*

2.09 **Matrices with multiple roots.** We shall now extend the main results of §2.07 to matrices whose roots are not necessarily simple. Suppose in the first place that x has only one distinct root and that its reduced characteristic function is $\varphi(\lambda) = (\lambda - \lambda_1)^\nu$, and set

$$\eta_1^i = \eta_i = (x - \lambda_1)^i = (x - \lambda_1)\eta_{i-1} \qquad (i = 1, 2, \cdots, \nu - 1);$$

then

$$\eta_1^\nu = 0, \quad x\eta_{\nu-1} = \lambda_1\eta_{\nu-1}, \quad x\eta_i = \lambda_1\eta_i + \eta_{i+1} \quad (i = 1, 2, \cdots, \nu - 2)$$

and

$$x^s = (\lambda_1 + \eta_1)^s = \lambda_1^s + s\lambda_1^{s-1}\eta_1 + \binom{s}{2}\lambda_1^{s-2}\eta_1^2 + \cdots$$

where the binomial expansion is cut short with the term $\eta_1^{\nu-1}$ since $\eta_1^\nu = 0$. Again, if $g(\lambda)$ is any scalar polynomial, then

$$g(x) = g(\lambda_1 + \eta_1) = g(\lambda_1) + g'(\lambda_1)\eta_1 + \cdots + \frac{g^{(\nu-1)}(\lambda_1)}{(\nu-1)!}\eta_1^{\nu-1}.$$

It follows immediately that, if $g^{(s)}(\lambda)$ is the first derivative of $g(\lambda)$ which is not 0 when $\lambda = \lambda_1$ and $(\kappa - 1)s < \nu \leq \kappa s$, then the reduced equation of $g(x)$ is

$$[g(x) - g(\lambda_1)]^\kappa = 0.$$

It should be noted that the first $\nu - 1$ powers of η_1 are linearly independent since $\varphi(\lambda)$ is the reduced characteristic function of x.

2.10 We shall now suppose that x has more than one root. Let the reduced characteristic function be

(21) $$\varphi(\lambda) = \prod_{i=1}^{r} (\lambda - \lambda_i)^{\nu_i} \qquad (\Sigma\nu_i = \nu, r > 1)$$

and set

(22) $$h_i(\lambda) = \varphi(\lambda)/(\lambda - \lambda_i)^{\nu_i}.$$

We can determine two scalar polynomials, $M_i(\lambda)$ and $N_i(\lambda)$, of degrees not exceeding $\nu_i - 1$ and $\nu - \nu_i - 1$, respectively, such that

$$M_i(\lambda)h_i(\lambda) + (\lambda - \lambda_i)^{\nu_i}N_i(\lambda) \equiv 1, \qquad M_i(\lambda_i) \neq 0.$$

If we set

(23) $$\varphi_i(\lambda) = M_i(\lambda)h_i(\lambda),$$

then $1 - \Sigma\varphi_i(\lambda)$ is exactly divisible by $\varphi(\lambda)$ and, being of degree $\nu - 1$ at most, must be identically 0; hence

(24) $$\sum_1^r \varphi_i(\lambda) = 1.$$

Again, from (22) and (23), $\varphi(\lambda)$ is a factor of $\varphi_i(\lambda)\varphi_j(\lambda)$ $(i \neq j)$ and hence on multiplying (24) by $\varphi_i(\lambda)$ we have

(25) $$[\varphi_i(\lambda)]^2 \equiv \varphi_i(\lambda), \qquad \varphi_i(\lambda)\varphi_j(\lambda) \equiv 0, \bmod \varphi(\lambda) \qquad (i \neq j).$$

Further, if $g(\lambda)$ is a scalar polynomial, then

(26) $$\begin{aligned} g(\lambda) &= \sum_1^r g(\lambda)\varphi_i(\lambda) \\ &= \sum_1^r \left[g(\lambda_i) + g'(\lambda_i)(\lambda - \lambda_i) + \cdots + \frac{g^{(\nu_i-1)}(\lambda_i)}{(\nu_i - 1)!}(\lambda - \lambda_i)^{\nu_i-1}\right]\varphi_i(\lambda) + R \end{aligned}$$

where R has the form $\Sigma C_i(\lambda)(\lambda - \lambda_i)^{\nu_i}\varphi_i(\lambda)$, C_i being a polynomial, so that R vanishes when x is substituted for λ.

2.11 If we put x for λ in (23) and set φ_i for $\varphi_i(x)$, then (24) and (25) show that

(27) $$\varphi_i^2 = \varphi_i, \quad \varphi_i\varphi_j = 0 \quad (i \neq j), \quad \sum_1^r \varphi_i = 1.$$

It follows as in §2.07 that the matrices φ_i are linearly independent and none is zero, since $\varphi_i(\lambda_i) \neq 0$ so that $\varphi(\lambda)$ is not a factor of $\varphi_i(\lambda)$, which would be the case were $\varphi_i(x) = 0$. We now put x for λ in (26) and set

(28) $$\eta_i = (x - \lambda_i)\varphi_i \qquad (i = 1, 2, \cdots, r).$$

Since the ν_ith power of $(\lambda - \lambda_i)\varphi_i(\lambda)$ is the first which has $\varphi(\lambda)$ as a factor, η_i is a nilpotent matrix of index ν_i (cf. §1.05) and, remembering that $\varphi_i^2 = \varphi_i$, we have

(29) $$\eta_i^j = (x - \lambda_i)^j\varphi_i \neq 0 \quad (j < \nu_i), \qquad \eta_i\varphi_i = \eta_i = \varphi_i\eta_i,$$

(30) $$x\varphi_i = \lambda_i\varphi_i + \eta_i, \quad x\eta_i^j = \lambda_i\eta_i^j + \eta_i^{j+1},$$

equation (26) therefore becomes

(31) $$g(x) = \sum_1^r \left[g(\lambda_i)\varphi_i + g'(\lambda_i)\eta_i + \cdots + \frac{g^{(\nu_i - 1)}(\lambda_i)}{(\nu_i - 1)!}\eta_i^{\nu_i - 1} \right]$$

and in particular

(32) $$x = \sum_1^r (\lambda_i\varphi_i + \eta_i) = \Sigma x_i.$$

The matrices φ_i and η_i are called the *principal idempotent* and *nilpotent elements* of x corresponding to the root λ_i. The matrices φ_i are uniquely determined by the following conditions: if ψ_i $(i = 1, 2, \cdots, r)$ are any matrices such that

(33)
(i) $x\psi_i = \psi_i x$,
(ii) $(x - \lambda_i)\psi_i$ is nilpotent,
(iii) $\sum_i \psi_i = 1, \qquad \psi_i^2 = \psi_i \neq 0$,

then $\psi_i = \varphi_i$ $(i = 1, 2, \cdots, r)$. For let $\theta_{ij} = \varphi_i\psi_j$; from (i) θ_{ij} also equals $\psi_j\varphi_i$. From (ii) and (28)

$$\eta_i = x\varphi_i - \lambda_i\varphi_i, \qquad \xi_j = x\psi_j - \lambda_j\psi_j$$

are both nilpotent and, since η_i and φ_i are polynomials in x, they are commutative with ψ_j and therefore with ξ_j; also

$$x\theta_{ij} = \lambda_i\theta_{ij} + (x - \lambda_i)\varphi_i\psi_j = \lambda_i\theta_{ij} + \eta_i\psi_j$$
$$= \lambda_j\theta_{ij} + (x - \lambda_j)\varphi_i\psi_j = \lambda_j\theta_{ij} + \xi_j\varphi_i.$$

Hence $(\lambda_i - \lambda_j)\theta_{ij} = \xi_j\varphi_i - \eta_i\psi_j$. But if μ is the greater of the indices of ξ_j and η_i, then, since all the matrices concerned are commutative, each term of $(\xi_j\varphi_i - \eta_i\psi_j)^{2\mu}$ contains ξ_j^μ or η_i^μ as a factor and is therefore 0. If $\theta_{ij} \neq 0$, this is impossible when $i \neq j$ since θ_{ij} is idempotent and $\lambda_i - \lambda_j \neq 0$. Hence $\varphi_i\psi_j = 0$ when $i \neq j$ and from (iii)

$$\psi_j = \psi_j\Sigma\varphi_i = \psi_j\varphi_j = \varphi_j\Sigma\psi_i = \varphi_j$$

which proves the uniqueness of the φ's.

2.12 We shall now determine the reduced equation of $g(x)$. If we set g_i for $g(x)\varphi_i$, then

$$\begin{aligned} g_i &= g(\lambda_i)\varphi_i + g'(\lambda_i)\eta_i + \cdots + \frac{g^{(\nu_i - 1)}(\lambda_i)}{(\nu_i - 1)!}\eta_i^{\nu_i - 1} \\ &= g(\lambda_i)\varphi_i + \zeta_i, \end{aligned} \tag{34}$$

say, and if s_i is the order of the first derivative in (34) which is not 0, then ζ_i is a nilpotent matrix whose index k_i is given by $k_i = 1 < \nu_i/s_i \leq k_i$.

If $\Phi(\lambda)$ is a scalar polynomial, and $\gamma_i = g(\lambda_i)$,

$$\Phi(g(x)) = \sum_i \Phi(g_i)\varphi_i = \sum_i \left[\Phi(\gamma_i)\varphi_i + \Phi'(\gamma_i)\zeta_i + \cdots + \frac{\Phi^{(k_i - 1)}(\gamma_i)}{(k_i - 1)!}\zeta_i^{k_i - 1}\right]$$

so that $\Phi(g(x)) = 0$ if, and only if, $g(\lambda_i)$ is a root of $\Phi(\lambda)$ of multiplicity k_i. Hence, if

$$\Psi(\lambda) = \Pi[\lambda - g(\lambda_i)]^{k_i}$$

where when two or more values of i give the same value of $g(\lambda_i)$, only that one is to be taken for which k_i is greatest, then $\Psi(\lambda)$ is the reduced characteristic function of $g(x)$. As a part of this result we have the following theorem.

THEOREM 5. *If $g(\lambda)$ is a scalar polynomial and x a matrix whose distinct roots are $\lambda_1, \lambda_2, \cdots, \lambda_r$, the roots of the matrix $g(x)$ are*[5]

$$g(\lambda_1), g(\lambda_2), \cdots, g(\lambda_r).$$

If the roots $g(\lambda_i)$ are all distinct, the principal idempotent elements of $g(x)$ are the same as those of x; for condition (33) of §2.11 as applied to $g(x)$ are satisfied by φ_i $(i = 1, 2, \cdots, r)$, and these conditions were shown to characterize the principal idempotent elements completely.

2.13 **The square root of a matrix.** Although the general question of functions of a matrix will not be taken up till a later chapter, it is convenient to give here one determination of the square root of a matrix x.

[5] That these are roots of $g(x)$ follows immediately from the fact that $\lambda - x$ is a factor of $g(\lambda) - g(x)$; but it does not follow so readily from this that the only roots are those given except, of course, when $r = n$ and all the quantities $g(\lambda_i)$ are distinct.

If α and β are scalars, $\alpha \neq 0$, and $(\alpha + \beta)^{\frac{1}{2}}$ is expanded formally in a Taylor series,

$$(\alpha + \beta)^{\frac{1}{2}} = \alpha^{\frac{1}{2}} \sum_0^{\infty} \delta_r \left(\frac{\beta}{\alpha}\right)^r$$

then, if $S_\nu = \alpha^{\frac{1}{2}} \sum_0^{\nu - 1} \delta_r(\beta/\alpha)^r$, it follows that

$$S_\nu^2 = \alpha + \beta + \alpha T_\nu \tag{35}$$

where T_ν is a polynomial in β/α which contains no power of β/α lower than the νth. If a and b are commutative matrices and a is the square of a known non-singular matrix $a^{\frac{1}{2}}$, then (35) being an algebraic identity in α and β remains true when a and b are put in their place.

If $x_i = \lambda_i \varphi_i + \eta_i$ is the matrix defined in §2.11 (32), then so long as $\lambda_i \neq 0$, we may set $\alpha = \lambda_i \varphi_i$, $\beta = \eta_i$ since $\lambda_i \varphi_i = (\lambda_i^{\frac{1}{2}} \varphi_i)^2$; and in this case the Taylor series terminates since $\eta_i^{\nu_i} = 0$, that is, $T_{\nu_i} = 0$ and the square of the terminating series for $(\lambda_i \varphi_i + \eta_i)^{\frac{1}{2}}$ in powers of η_i equals $\lambda_i \varphi_i + \eta_i$. It follows immediately from (32) and (27) that, if x is a matrix no one of whose roots is 0, the square of the matrix

$$x^{\frac{1}{2}} = \sum_1^r \lambda_i^{\frac{1}{2}} \left[\varphi_i + \tfrac{1}{2} \lambda_i^{-1} \eta_i - \cdots + (-1)^{\nu_i - 2} \frac{(2\nu_i - 4)!}{2^{2\nu_i - 3}(\nu_i - 2)!(\nu_i - 1)!} \left(\frac{\eta_i}{\lambda_i}\right)^{\nu_i - 1} \right] \tag{36}$$

is x.

If the reduced equation of x has no multiple roots, (36) becomes

$$x^{\frac{1}{2}} = \Sigma \lambda_i^{\frac{1}{2}} \varphi_i \tag{37}$$

and this is valid even if one of the roots is 0. If, however, 0 is a multiple root of the reduced equation, x may have no square root as, for example, the matrix $\begin{Vmatrix} 0 & 1 \\ 0 & 0 \end{Vmatrix}$.

Formula (36) gives 2^r determinations of $x^{\frac{1}{2}}$ but we shall see later that an infinity of determinations is possible in certain cases.

2.14 **Reducible matrices.** If $x = x_1 + x_2$ is the direct sum of x_1 and x_2 and e_1, e_2 are the corresponding idempotent elements, that is,

$$e_i x = x_i = x_i, \qquad e_i e_j = 0 \qquad (i \neq j;\, i, j = 1, 2),$$

then $x^r = x_1^r + x_2^r$ $(r \geq 2)$ and we may set as before $1 = x^0 = x_1^0 + x_2^0 = e_1 + e_2$. Hence, if $f(\lambda) = \lambda^m + b_1 \lambda^{m-1} + \cdots + b_m$ is any scalar polynomial, we have

$$f(x) = e_1 f(x_1) + e_2 f(x_2) = f(x_1) + f(x_2) - b_m,$$

and if $g(\lambda)$ is a second scalar polynomial

$$f(x)g(x) = e_1f(x_1)g(x_1) + e_2f(x_2)g(x_2).$$

Now if $f_i(\lambda)$ is the reduced characteristic function of x_i regarded as a matrix in the space determined by e_i, then the reduced characteristic function of x_i as a matrix in the original fundamental space is clearly $\lambda f_i(\lambda)$ unless λ is a factor of $f_i(\lambda)$ in which case it is simply $f_i(\lambda)$. Further the reduced characteristic function of $x = x_1 + x_2$ is clearly the least common multiple of $f_1(\lambda)$ and $f_2(\lambda)$; for if

$$\psi(\lambda) = f_1(\lambda)g_1(\lambda) = f_2(\lambda)g_2(\lambda)$$

then

$$\begin{aligned}\psi(x_1 + x_2) &= e_1\psi(x_1) + e_2\psi(x_2)\\ &= e_1f_1(x_1)g_1(x_1) + e_2f_2(x_2)g_2(x_2) = 0.\end{aligned}$$

CHAPTER III

INVARIANT FACTORS AND ELEMENTARY DIVISORS

3.01 **Elementary transformations.** By an elementary transformation of a matric polynomial $a(\lambda) = \| a_{ij} \|$ is meant one of the following operations on the rows or columns.

Type I. The operation of adding to a row (column) a different row (column) multiplied by a scalar polynomial $\theta(\lambda)$.

Type II. The operation of interchanging two rows (columns).

Type III. The operation of multiplying a row (column) by a constant $k \neq 0$.

These transformations can be performed algebraically as follows.

Type I. Let

$$P_{ij} = 1 + \theta(\lambda)e_{ij} \quad (i \neq j),$$

$\theta(\lambda)$ being a scalar polynomial; then $| P_{ij} | = 1$ and

$$P_{ij}a = \sum_{p,q} a_{pq}e_{pq} + \theta \sum_{q} a_{jq}e_{iq}$$

which is the matrix derived from $a(\lambda)$ by adding θ times the jth row to the ith. The corresponding operation on the columns is equivalent to forming the product aP_{ji}.

Type II. Let Q_{ij} be the matrix

$$Q_{ij} = 1 - e_{ii} - e_{jj} + e_{ij} + e_{ji} \qquad (i \neq j)$$

that is, Q_{ij} is the matrix derived from the identity matrix by inserting 1 in place of 0 in the coefficients of e_{ij} and e_{ji} and 0 in place of 1 in the coefficients of e_{ii} and e_{jj}; then $| Q_{ij} | = -1$ and

$$Q_{ij}a = \sum_{p,q} a_{pq}e_{pq} - \sum_{q} a_{iq}e_{iq} - \sum_{q} a_{jq}e_{jq} + \sum_{q} a_{jq}e_{iq} + \sum_{q} a_{iq}e_{jq},$$

that is, $Q_{ij}a$ is derived from a by interchanging the ith and jth rows. Similarly aQ_{ij} is obtained by interchanging the ith and jth columns.

Since any permutation can be effected by a succession of transpositions, the corresponding transformation in the rows (columns) of a matrix can be produced by a succession[1] of transformations of Type II.

Type III. This transformation is effected on the rth row (column) by multiplying on the left (right) by $R = 1 + (k - 1)e_{rr}$; it is used only when it is convenient to make the leading coefficient in some term equal to 1.

[1] The transformation corresponding to the substitution $\begin{pmatrix} 1 & 2 & \dots & n \\ p_1 & p_2 & \dots & p_n \end{pmatrix}$ is $Q = \sum_i e_{ip_i}$.

The inverses of the matrices used in these transformations are

$$P_{ij}^{-1} = 1 - \theta e_{ij}, \quad Q_{ij}^{-1} = Q_{ij}, \quad R^{-1} = 1 + (k^{-1} - 1)e_{rr};$$

these inverses are elementary transformations. The transverses are also elementary since $P'_{ij} = P_{ji}$, and Q_{ij} and R are symmetric.[2]

A matric polynomial $b(\lambda)$ which is derived from $a(\lambda)$ by a sequence of elementary transformations is said to be *equivalent* to $a(\lambda)$; every such polynomial has the form $p(\lambda)a(\lambda)q(\lambda)$ where p and q are products of elementary transformations. Since the inverse of an elementary transformation is elementary, $a(\lambda)$ is also equivalent to $b(\lambda)$. Further, the inverses of p and q are polynomials so that these are what we have already called elementary polynomials; we shall see later that every elementary polynomial can be derived from 1 by a sequence of elementary transformations.

In the following sections we require two lemmas whose proofs are almost immediate.

LEMMA 1. *The rank*[3] *of a matrix is not altered by an elementary transformation.*

For if $|P| \neq 0$, AP and PA have the same rank as A (§1.10).

LEMMA 2. *The highest common factor of the coordinates of a matric polynomial is not altered by an elementary transformation.*

This follows immediately from the definition of elementary transformations.

3.02 The normal form of a matrix. The theorem we shall prove in this section is as follows.

THEOREM 1. *If $a(\lambda)$ is a matric polynomial of rank r, it can be reduced by elementary transformations to a diagonal matrix*

$$(1) \qquad \sum_1^r \alpha_i(\lambda)e_{ii} = \left\| \begin{matrix} \alpha_1(\lambda) & & & & & & \\ & \alpha_2(\lambda) & & & & & \\ & & \ddots & & & & \\ & & & \alpha_r(\lambda) & & & \\ & & & & 0 & & \\ & & & & & \ddots & \\ & & & & & & 0 \end{matrix} \right\| = P(\lambda)a(\lambda)Q(\lambda),$$

[2] The definition of an elementary transformation given above is the most convenient but not the only possible one. All three transformations have the form $T = 1 + xSy$ with the condition that $1 + Sxy$ is not 0 and is independent of λ.

[3] By the rank of a matric polynomial is meant the order of the highest minor which does not vanish identically. For particular values of λ the rank may be smaller than r; there are always values of λ for which it equals r and it cannot be greater.

where the coefficient of the highest power of λ *in each polynomial* $\alpha_i(\lambda)$ *is* 1, α_i *is a factor of* $\alpha_{i+1}, \cdots, \alpha_r$ $(i = 1, 2, \cdots, r - 1)$, *and* $P(\lambda)$, $Q(\lambda)$ *are elementary polynomials.*

We shall first show that, if the coordinate of $a(\lambda)$ of minimum degree m, say a_{pq}, is not a factor of every other coordinate, then $a(\lambda)$ is equivalent to a matrix in which the degree of the coordinate of minimum degree is less than m.

Suppose that a_{pq} is not a factor of a_{pi} for some i; then we may set $a_{pi} = \beta a_{pq} + a'_{pi}$ where β is integral and a'_{pi} is not 0 and is of lower degree than m. Subtracting β times the qth column from the ith we have an equivalent matrix in which the coordinate[4] (p, i) is a'_{pi} whose degree is less than m. The same reasoning applies if a_{pq} is not a factor of every coordinate a_{iq} in the qth column.

After a finite number of such steps we arrive at a matrix in which a coordinate of minimum degree, say k_{pq}, is a factor of all the coordinates which lie in the same row or column, but is possibly not a factor of some other coordinate k_{ij}. When this is so, let $k_{pj} = \beta k_{pq}$, $k_{iq} = \gamma k_{pq}$ where β and γ are integral. If we now add $(1 - \beta)$ times the qth column to the jth, (p, j) and (i, j) become respectively

$$k'_{pj} = k_{pj} + (1 - \beta)k_{pq} = k_{pq},\ k'_{ij} = k_{ij} + (1 - \beta)k_{iq} = k_{ij} + (1 - \beta)\gamma k_{pq}.$$

Here either the degree of k'_{ij} is less than that of k_{pq}, or k'_{pj} has the minimum degree and is not a factor of k'_{ij}, which lies in the same column, and hence the minimum degree can be lowered as above.

The process just described can be repeated so long as the coordinate of lowest degree is not a factor of every other coordinate and, since each step lowers the minimum degree, we derive in a finite number of steps a matrix $\| b'_{ij} \|$ which is equivalent to $a(\lambda)$ and in which the coordinate of minimum degree is in fact a divisor of every other coordinate; and further we may suppose that $b'_{11} = \alpha_1(\lambda)$ is a coordinate of minimum degree and set $b'_{1i} = \gamma_i b'_{11}$, $b'_{j1} = \delta_j b'_{11}$. Subtracting γ_i times the first column from the ith and then δ_j times the first row from the jth $(i, j = 2, 3, \cdots, n)$ all the coordinates in the first row and column except b'_{11} become 0, and we have an equivalent matrix in the form

$$\text{(2)} \qquad \begin{matrix} \alpha_1(\lambda) & 0 & 0 & \cdots & 0 \\ 0 & b_{22} & b_{23} & \cdots & b_{2n} \\ 0 & b_{32} & b_{33} & \cdots & b_{3n} \\ \cdot & \cdot & \cdot & \cdots & \cdot \\ 0 & b_{n2} & b_{n3} & \cdots & b_{nn} \end{matrix}$$

in which α_1 is a factor of every b_{ij}. The coefficient of the highest power of λ in α_1 may be made 1 by a transformation of type III.

The theorem now follows readily by induction. For, assuming it is true for

[4] That is, the coordinate in the pth row and ith column.

matrices of order $n - 1$, the matrix of this order formed by the b's in (2) can be reduced to the diagonal matrix

$$\begin{matrix} \alpha_2(\lambda) & & & & & & \\ & \alpha_3(\lambda) & & & & & \\ & & \ddots & & & & \\ & & & \alpha_s(\lambda) & & & \\ & & & & 0 & & \\ & & & & & \ddots & \\ & & & & & & 0 \end{matrix}$$

where the α's satisfy the conditions of the theorem and each has α_1 as a factor (§3.01, Lemma 2). Moreover, the elementary transformations by which this reduction is carried out correspond to transformations affecting the last $n - 1$ rows and columns alone in (2) and, because of the zeros in the first row and column, these transformations when applied to (2) do not affect its first row and column; also, since elementary transformations do not affect the rank (§3.01, Lemma 1), s equals r and $a(\lambda)$ has therefore been reduced to the form required by the theorem.

The theorem is clearly true for matrices of order 1 and hence is true for any order.

Corollary. A matric polynomial whose determinant is independent of λ and is not 0, that is, an elementary polynomial, can be derived from 1 by the product of a finite number of elementary transformations.

The polynomials α_i are called the *invariant factors* of $a(\lambda)$.

3.03 **Determinantal and invariant factors.** The *determinantal factor* of the sth order, D_s, of a matric polynomial $a(\lambda)$ is defined as the highest common factor of all minors of order s, the coefficient of the highest power of λ being taken as 1. An elementary transformation of type I either leaves a given minor unaltered or changes it into the sum of that minor and a multiple of another of the same order, and a transformation of type II simply permutes the minors of a given order among themselves, while one of type III merely multiplies a minor by a constant different from 0. Hence equivalent matrices have the same determinantal factors. Bearing this in mind we see immediately from the form of (1) that the determinantal factors of $a(\lambda)$ are given by

$$D_s = \alpha_1\alpha_2 \cdots \alpha_s \qquad (s = 1, 2, \cdots, r), \qquad D_s = 0 \ (s > r),$$

so that[5]

$$\alpha_s = D_s/D_{s-1}.$$

The invariant factors are therefore known when the determinantal factors are given, and vice versa.

[5] Since α_{s-1} is a factor of α_s, it follows that also D_s^2 is a factor of $D_{s-1}D_{s+1}$.

The definitions of this and the preceding sections have all been made relative to the fundamental basis. But we have seen in §1.08 that, if a_1 is the matrix with the same array of coordinates as a but relative to another basis, then there exists a non-singular constant matrix b such that $a = b^{-1}a_1b$ so that a and a_1 are equivalent matrices. In terms of the new basis a_1 has the same invariant factors as a does in terms of the old and a, being equivalent to a_1, has therefore the same invariant factors in terms of the new basis as it has in the old. Hence the invariant and determinantal factors of a matric polynomial are independent of the (constant) basis in terms of which its coordinates are expressed.

The results of this section may be summarized as follows.

THEOREM 2. *Two matric polynomials are equivalent if, and only if, they have the same invariant factors.*

3.04 **Non-singular linear polynomials.** In the case of linear polynomials Theorem 2 can be made more precise as follows.

THEOREM 3. *If $a\lambda + b$ and $c\lambda + d$ are non-singular linear polynomials which have the same invariant factors, and if $|c| \neq 0$, there exist non-singular constant matrices p and q such that*

$$p(a\lambda + b)q = c\lambda + d.$$

We have seen in Theorem 2 that there exist elementary polynomials $P(\lambda)$, $Q(\lambda)$ such that

$$c\lambda + d = P(\lambda)(a\lambda + b)Q(\lambda). \tag{3}$$

Since $|c| \neq 0$, we can employ the division transformation to find matric polynomials p_1, q_1 and constant matrices p, q such that

$$P(\lambda) = (c\lambda + d)p_1 + p, \quad Q(\lambda) = q_1(c\lambda + d) + q.$$

Using this in (3) we have

$$\begin{aligned} c\lambda + d = p(a\lambda + b)q + (c\lambda + d)p_1(a\lambda + b)Q + P(a\lambda + b)q_1(c\lambda + d) \\ - (c\lambda + d)p_1(a\lambda + b)q_1(c\lambda + d) \end{aligned} \tag{4}$$

and, since from (3)

$$(a\lambda + b)Q = P^{-1}(c\lambda + d), \quad P(a\lambda + b) = (c\lambda + d)Q^{-1},$$

we may write in place of (4)

$$\begin{aligned} p(a\lambda + b)q &= [1 - (c\lambda + d)(p_1P^{-1} + Q^{-1}q_1 - p_1(a\lambda + b)q_1)](c\lambda + d) \\ &= [1 - (c\lambda + d)R](c\lambda + d) \end{aligned} \tag{5}$$

where $R = pP_1^{-1} + Q^{-1}q_1 - p_1(a\lambda + b)q_1$, which is integral in λ since P and Q are elementary. If $R \neq 0$, then, since $|c| \neq 0$, the degree of the right side of (5) is at least 2, whereas the degree of the left side is only 1; hence $R = 0$ so that (5) gives $p(a\lambda + b)q = c\lambda + d$. Since $c\lambda + d$ is not singular, neither p nor q can be singular, and hence the theorem is proved.

When $|c| = 0$ (and therefore also $|a| = 0$) the remaining conditions of Theorem 3 are not sufficient to ensure that we can find constant matrices in place of P and Q, but these conditions are readily modified so as to apply to this case also. If we replace λ by λ/μ and then multiply by μ, $a\lambda + b$ is replaced by the homogeneous polynomial $a\lambda + b\mu$; and the definition of invariant factors applies immediately to such polynomials. In fact, if $|a| \neq 0$, the invariant factors of $a\lambda + b\mu$ are simply the homogeneous polynomials which are equivalent to the corresponding invariant factors of $a\lambda + b$. If, however, $|a| = 0$, then $|a\lambda + b\mu|$ is divisible by a power of μ which leads to factors of the form μ^i in the invariant factors of $a\lambda + b\mu$ which have no counterpart in those of $a\lambda + b$.

If $|c| = 0$ but $|c\lambda + d| \not\equiv 0$, there exist values, $\lambda_1 \neq 0$, μ_1, such that $|c\lambda_1 + d\mu_1| \neq 0$ and, if we make the transformation

$$(6) \qquad \lambda = \lambda_1\alpha, \quad \mu = \mu_1\alpha + \beta,$$

$a\lambda + b\mu$, $c\lambda + d\mu$ become $a_1\alpha + b_1\beta$, $c_1\alpha + d_1\beta$ where $a_1 = a\lambda_1 + b\mu_1$, $c_1 = c\lambda_1 + d\mu_1$, and therefore $|c_1| \neq 0$. Further, when $a\lambda + b\mu$ and $c\lambda + d\mu$ have the same invariant factors, this is also true of $a_1\alpha + b_1\beta$ and $c_1\alpha + d_1\beta$. Since $|c_1| \neq 0$, the proof of Theorem 3 is applicable, so that there are constant non-singular matrices p, q for which $p(a_1\alpha + b_1\beta)q = c_1\alpha + d_1\beta$, and on reversing the substitution (6) we have

$$p(a\lambda + b\mu)q = c\lambda + d\mu.$$

Theorem 3 can therefore be extended as follows.

THEOREM 4. *If the non-singular polynomials $a\lambda + b\mu$, $c\lambda + d\mu$ have the same invariant factors, there exist non-singular constant matrices p, q such that $p(a\lambda + b\mu)q = c\lambda + d\mu$.*

An important particular case of Theorem 3 arises when the polynomials have the form $\lambda - b$, $\lambda - d$. For if $p(\lambda - b)q = \lambda - d$, on equating coefficients we have $pq = 1$, $pbq = d$; hence $b = p^{-1}dp$, that is, b and d are similar. Conversely, if b and d are similar, then $\lambda - b$ and $\lambda - d$ are equivalent, and hence we have the following theorem.

THEOREM 5. *Two constant matrices b, d are similar if, and only if, $\lambda - b$ and $\lambda - d$ have the same invariant factors.*

3.05 Elementary divisors. If $D = |a\lambda + b|$ is not identically zero and if $\lambda_1, \lambda_2, \cdots, \lambda_s$ are its distinct roots, say

$$D = (\lambda - \lambda_1)^{\nu_1}(\lambda - \lambda_2)^{\nu_2} \cdots (\lambda - \lambda_s)^{\nu_s},$$

then the invariant factors of $a\lambda + b$, being factors of D, have the form

$$(7) \qquad \begin{aligned} \alpha_1 &= (\lambda - \lambda_1)^{\nu_{11}}(\lambda - \lambda_2)^{\nu_{12}} \cdots (\lambda - \lambda_s)^{\nu_{1s}} \\ \alpha_2 &= (\lambda - \lambda_1)^{\nu_{21}}(\lambda - \lambda_2)^{\nu_{22}} \cdots (\lambda - \lambda_s)^{\nu_{2s}} \\ &\cdots\cdots\cdots\cdots\cdots\cdots\cdots\cdots \\ \alpha_i &= (\lambda - \lambda_1)^{\nu_{i1}}(\lambda - \lambda_2)^{\nu_{i2}} \cdots (\lambda - \lambda_s)^{\nu_{is}} \\ &\cdots\cdots\cdots\cdots\cdots\cdots\cdots\cdots \\ \alpha_n &= (\lambda - \lambda_1)^{\nu_{n1}}(\lambda - \lambda_2)^{\nu_{n2}} \cdots (\lambda - \lambda_s)^{\nu_{ns}} \end{aligned}$$

where $\sum_{j=1}^{n} \nu_{ji} = \nu_i$ and, since α_j is a factor of α_{j+1},

$$(8) \qquad \nu_{1i} \leq \nu_{2i} \leq \cdots \leq \nu_{ni} \; (i = 1, 2, \cdots, s).$$

Such of the factors $(\lambda - \lambda_j)^{\nu_{ij}}$ as are not constants, that is, those for which $\nu_{ij} > 0$, are called the *elementary divisors* of $a\lambda + b$. The elementary divisors of $\lambda - b$ are also called the elementary divisors of b. When all the exponents ν_{ij} which are not 0 equal 1, b is said to have *simple* elementary divisors.

For some purposes the degrees of the elementary divisors are of more importance than the divisors themselves and, when this is the case, they are indicated by writing

$$(9) \qquad [(\nu_{n1}, \nu_{n-1,1}, \cdots, \nu_{11}), (\nu_{n2}, \nu_{n-1,2}, \cdots, \nu_{12}), \cdots]$$

where exponents belonging to the same linear factor are in the same parenthesis, zero exponents being omitted; (9) is sometimes called the characteristic of $a\lambda + b$. If a root, say λ_1, is zero, it is convenient to indicate this by writing ν_{i1}^0 in place of ν_{i1}.

The maximum degree of $|a\lambda + b|$ is n and therefore $\sum_{i,j} \nu_{ij} \leq n$ where the equality sign holds only when $|a| \neq 0$.

The modifications necessary when the homogeneous polynomial $a\lambda + b\mu$ is taken in place of $a\lambda + b$ are obvious and are left to the reader.

3.06 Matrices with given elementary divisors. The direct investigation of the form of a matrix with given elementary divisors is somewhat tedious. It can be carried out in a variety of ways; but, since the form once found is easily verified, we shall here state this form and give the verification, merely saying in passing that it is suggested by the results of §2.07 together with a study of a matrix whose reduced characteristic function is $(\lambda - \lambda_1)^\nu$.

THEOREM 6. *If $\lambda_1, \lambda_2, \cdots, \lambda_s$ are any constants, not necessarily all different and $\nu_1, \nu_2, \cdots, \nu_s$ are positive integers whose sum is n, and if a_i is the array of ν_i rows and columns given by*

$$(10) \qquad \begin{matrix} \lambda_i & 1 & 0 & \cdots & 0 & 0 \\ 0 & \lambda_i & 1 & \cdots & 0 & 0 \\ \cdot & \cdot & \cdot & \cdots & \cdot & \cdot \\ \cdot & \cdot & \cdot & \cdots & \cdot & \cdot \\ 0 & 0 & 0 & \cdots & \lambda_i & 1 \\ 0 & 0 & 0 & \cdots & 0 & \lambda_i \end{matrix}$$

where each coordinate on the main diagonal equals λ_i, *those on the parallel on its right are* 1, *and the remaining ones are* 0, *and if a is the matrix of n rows and columns given by*

$$a = \begin{Vmatrix} a_1 & & & \\ & a_2 & & \\ & & \ddots & \\ & & & a_s \end{Vmatrix} \tag{11}$$

composed of blocks of terms defined by (10) *arranged so that the main diagonal of each lies on the main diagonal of a, the other coordinates being* 0, *then* $\lambda - a$ *has the elementary divisors*

$$(\lambda - \lambda_1)^{\nu_1}, (\lambda - \lambda_2)^{\nu_2}, \cdots, (\lambda - \lambda_s)^{\nu_s} \tag{12}$$

In addition to using a_i to denote the block given in (10) we shall also use it for the matrix having this block in the position indicated in (11) and zeros elsewhere. In the same way, if f_i is a block with ν_i rows and columns with 1's in the main diagonal and zeros elsewhere, we may also use f_i for the corresponding matrix of order n. We can then write

$$\lambda - a = \Sigma(\lambda f_i - a_i), \quad f_i a = a_i = af_i, \quad \Sigma f_i = 1.$$

The block of terms corresponding to $\lambda f_i - a_i$ has then the form

$$\begin{matrix} \lambda - \lambda_i & -1 & & & \\ & \lambda - \lambda_i & -1 & & \\ & & \ddots & \ddots & \\ & & & \ddots & -1 \\ & & & & \lambda - \lambda_i \end{matrix} \qquad (\nu_i \text{ rows and columns}) \tag{13}$$

where only the non-zero terms are indicated. The determinant of these ν_i rows and columns is $(\lambda - \lambda_i)^{\nu_i}$ and this determinant has a first minor equal to ± 1; the invariant factors of $\lambda f_i - a_i$, regarded as a matrix of order ν_i, are therefore $1, 1, \cdots, 1, (\lambda - \lambda_i)^{\nu_i}$ and hence it can be reduced by elementary transformation to the diagonal form

$$\begin{matrix} (\lambda - \lambda_i)^{\nu_i} & & & \\ & 1 & & \\ & & \ddots & \\ & & & 1. \end{matrix}$$

If we apply the same elementary transformations to the corresponding rows and columns of $\lambda - a$, the effect is the same as regards the block of terms $\lambda f_i - a_i$ (corresponding to a_i in (11)) since all the other coordinates in the rows and columns which contain elements of this block are 0; moreover these transformations do not affect the remaining blocks $\lambda f_j - a_j$ $(j \neq i)$ nor any 0 coordinate. Carrying out this process for $i = 1, 2, \cdots, s$ and permuting rows and columns, if necessary, we arrive at the form

$$\begin{matrix} (\lambda - \lambda_1)^{\nu} & & & & & & \\ & (\lambda - \lambda_2)^{\nu_2} & & & & & \\ & & \ddots & & & & \\ & & & (\lambda - \lambda_s)^{\nu_s} & & & \\ & & & & 1 & & \\ & & & & & \ddots & \\ & & & & & & 1. \end{matrix}$$

Suppose now that the notation is so arranged that

$$\lambda_1 = \lambda_2 = \cdots = \lambda_p = \alpha, \quad \nu_1 \geq \nu_2 \geq \cdots \geq \nu_p,$$

but $\lambda_i \neq \alpha$ for $i > p$. The nth determinantal factor D_n then contains $(\lambda - \alpha)$ to the power $\sum_1^p \nu_i$ exactly. Each minor of order $n - 1$ contains at least $p - 1$ of the factors

$$(14) \qquad (\lambda - \alpha)^{\nu_1}, (\lambda - \alpha)^{\nu_2}, \cdots, (\lambda - \alpha)^{\nu_p}$$

and in one the highest power $(\lambda - \alpha)^{\nu}$ is lacking; hence D_{n-1} contains $(\lambda - \alpha)$ to exactly the power $\sum_2^p \nu_i$ and hence the nth invariant factor α_n contains it to exactly the ν_1th power. Similarly the minors of order $n - 2$ each contain at least $p - 2$ of the factors (14) and one lacks the two factors of highest degree; hence $(\lambda - \alpha)$ is contained in D_{n-2} to exactly the power $\sum_3^p \nu_i$ and in α_{n-1} to the power ν_2. Continuing in this way we see that (14) gives the elementary divisors of a which are powers of $(\lambda - \alpha)$ and, treating the other roots in the same way, we see that the complete list of elementary divisors is given by (12) as required by the theorem.

3.07 If A is a matrix with the same elementary divisors as a, it follows from Theorem 5 that there is a matrix P such that $A = PaP^{-1}$ and hence, if we choose in place of the fundamental basis $(e_1, e_2, \cdots, e_n)$ the basis $(Pe_1, Pe_2, \cdots, Pe_n)$, it follows from Theorem 6 of chapter 1 that (11) gives the form of A relative to the new basis. This form is called the canonical form of A. It follows immediately from this that

$$(15) \qquad P^{-1}A^kP = \left\| \begin{matrix} a_1^k & & & \\ & a_2^k & & \\ & & \ddots & \\ & & & a_s^k \end{matrix} \right\|$$

where a_i^k is the block of terms derived by forming the kth power of a_i regarded as a matrix of order ν_i.

Since D_n equals $|\lambda - a|$, it is the characteristic function of a (or A) and, since D_{n-1} is the highest common factor of the first minors, it follows from Theorem 3 of chapter 2 that α_n is the reduced characteristic function.

If we add the f's together in groups each group consisting of all the f's that correspond to the same value of λ_i, we get a set of idempotent matrices, say $\varphi_1, \varphi_2, \cdots, \varphi_r$, corresponding to the distinct roots of a, say $\alpha_1, \alpha_2, \cdots, \alpha_r$. These are the principal idempotent elements of a; for (i) $a\varphi_i = \varphi_i a$, (ii) $(a - \alpha_i)\varphi_i$ is nilpotent, (iii) $\Sigma\varphi_i = \Sigma f_i = 1$ and $\varphi_i\varphi_j = 0$ $(i \neq j)$ so that the conditions of §2.11 are satisfied.

When the same root α_i occurs in several elementary divisors, the corresponding f's are called *partial idempotent elements* of a; they are not unique as is seen immediately by taking $a = 1$.

If α is one of the roots of A, the form of $A - \alpha$ is sometimes important. Suppose that $\lambda_1 = \lambda_2 = \cdots = \lambda_p = \alpha$, $\lambda_i \neq \alpha$ $(i > p)$ and set

$$b_i = a_i - \alpha f_i,$$

the corresponding array in the ith block of $a - \alpha$ (cf. (10), (11)) being

$$(16) \qquad \begin{matrix} \lambda_i - \alpha & 1 & & \\ & \lambda_i - \alpha & 1 & \\ & & \ddots & \ddots \\ & & & \cdot \; 1 \\ & & & \lambda_i - \alpha. \end{matrix}$$

In the case of the first p blocks $\lambda_i - \alpha = 0$ and the corresponding $b_1, b_2, \cdots, b_p$ are nilpotent, the index of b_i being ν_i and, assuming $\nu_1 \geq \nu_2 > \cdots \geq \nu_p$ as before, $(A - \alpha)^k$ has the form

$$P^{-1}(A - \alpha)^k P = \begin{Vmatrix} b_1^k & & & \\ & b_2^k & & \\ & & \ddots & \\ & & & b_s^k \end{Vmatrix}$$

or, when $k \geq \nu_1$,

$$(17) \qquad P^{-1}(A - \alpha)^k P = \begin{Vmatrix} 0 & & & & & \\ & \ddots & & & & \\ & & 0 & & & \\ & & & b_{p+1}^k & & \\ & & & & \ddots & \\ & & & & & b_s^k \end{Vmatrix}$$

Since none of the diagonal coordinates of $b_{p+1}, \cdots, b_s$ are 0, the rank of $(A - \alpha)^k$, when $k \geq \nu_1$, is exactly $n - \sum_1^p \nu_i = \sum_{p+1}^s \nu_i$ and the nullspace of $(A - \alpha)^k$ is then the same as that of $(A - \alpha)^{\nu_1}$. Hence, if there exists a vector z such that $(A - \alpha)^k z = 0$ but $(A - \alpha)^{k-1} z \neq 0$, then (i) $k \leq \nu_1$, (ii) z lies in the nullspace of $(A - \alpha)^{\nu_1}$.

3.08 **Invariant vectors.** If A is a matrix with the elementary divisors given in the statement of Theorem 6, then $\lambda - A$ is equivalent to $\lambda - a$ and by Theorem 5 there is a non-singular matrix P such that $A = PaP^{-1}$. If we denote the unit vectors corresponding to the rows and columns of a_i in (10) by $e_1^i, e_2^i, \cdots, e_{\nu_i}^i$ and set

$$x_j^i = \begin{cases} Pe_j^i (j = 1, 2, \cdots, \nu_i; i = 1, 2, \cdots, s) \\ 0 \ (j < 1 \text{ or } > \nu_i, \text{ or } i < 1 \text{ or } > s) \end{cases} \tag{18}$$

then

$$ae_1^i = \lambda_i e_1^i, \ ae_2^i = \lambda_i e_2^i + e_1^i, \ \cdots, \ ae_{\nu_i}^i = \lambda_i e_{\nu_i}^i + e_{\nu_i - 1}^i$$

and hence

$$Ax_j^i = \lambda_i x_j^i + x_{j-1}^i \quad (j = 1, 2, \cdots, \nu_i; i = 1, 2, \cdots, s). \tag{19}$$

The vectors x_j^i are called a set of *invariant vectors*[6] of A.

The matrix A can be expressed in terms of its invariant vectors as follows. We have from (10)

$$a_i = \sum_j (\lambda_i e_j^i + e_{j-1}^i) S e_j^i = \sum_j e_j^i S(\lambda_i e_j^i + e_{j+1}^i)$$

and hence, if

$$y_j^i = (P')^{-1} e_j^i = (PP')^{-1} x_j^i, \tag{20}$$

then

$$A = \sum_{i,j} (\lambda_i x_j^i + x_{j-1}^i) S y_j^i = \sum_{i,j} x_j^i S(\lambda_i y_j^i + y_{j+1}^i) \tag{21}$$

where it should be noted that the y's form a system reciprocal to the x's and that each of these systems forms a basis of the vector space since $| P | \neq 0$.

If we form the transverse of A, we have from (21)

$$A' = \sum_{i,j} (\lambda_i y_j^i + y_{j+1}^i) S x_j^i \tag{22}$$

[6] If homogeneous coordinates are used so that vectors represent points, an invariant vector is usually called a pole.

so that the invariant vectors of A' are obtained by forming the system reciprocal to the x's and inverting the order in each group of vectors corresponding to a given elementary divisor; thus

$$A'y_{\nu_i}^i = \lambda_i y_{\nu_i}^i,\ A'y_{\nu_i - 1}^i = \lambda_i y_{\nu_i - 1}^i + y_{\nu_i}^i,\ \cdots,\ A'y_1^i = \lambda_i y_1^i + y_2^i.$$

A matrix A and its transverse clearly have the same elementary divisors and are therefore similar. The matrix which transforms A into A' can be given explicitly as follows. Let q_i be the symmetric array

$$\begin{matrix} 0 & 0 & \cdots & 0 & 1 \\ 0 & 0 & \cdots & 1 & 0 \\ \cdots & \cdots & \cdots & \cdots & \cdots \\ \cdots & \cdots & \cdots & \cdots & \cdots \\ 0 & 1 & \cdots & 0 & 0 \\ 1 & 0 & \cdots & 0 & 0 \end{matrix} \qquad (\nu_i \text{ rows and columns}).$$

It is easily seen that $q_i a_i = a_i' q_i$ and hence, if Q is the matrix

$$Q = \begin{Vmatrix} q_1 & & & & \\ & q_2 & & & \\ & & \cdot & & \\ & & & \cdot & \\ & & & & q_s \end{Vmatrix}$$

we have $Qa = a'Q$, and a short calculation gives $A' = R^{-1}AR$ where R is the symmetric matrix

$$R = PQ^{-1}P' = PQP'. \tag{23}$$

If the elementary divisors of A are simple, then $Q = 1$ and $R = PP'$.

If the roots λ_i of the elementary divisors (12) are all different, the nullity of $(A - \lambda_i)$ is 1, and hence x_1^i is unique to a scalar multiplier. But the remaining x_j^i are not unique. In fact, if the x's denote one choice of the invariant vectors, we may take in place of x_j^i

$$z_j^i = k_1^i x_j^i + k_2^i x_{j-1}^i + \cdots + k_j^i x_1^i \quad (j = 1, 2, \cdots, \nu_i)$$

where the k's are any constant scalars subject to the condition $k_1^i \neq 0$. Suppose now that $\lambda_1 = \lambda_2 = \cdots = \lambda_p = \alpha$, $\lambda_i \neq \alpha$ $(i > p)$ and $\nu_1 \geq \nu_2 \geq \cdots \geq \nu_p$ as in §3.07. We shall say that $z_1, z_2, \cdots, z_k$ is a *chain*[7] of invariant vectors *belonging to the exponent* k if

$$\begin{aligned} z_i &= (A - \alpha)^{k-i} z_k \neq 0 \qquad (i = 1, 2, \cdots, k) \\ (A - \alpha)^k z_k &= 0. \end{aligned} \tag{24}$$

It is also convenient to set $z_i = 0$ for $i < 0$ or $> k$. We have already seen that $k \leq \nu_1$ and that z_k lies in the nullspace of $(A - \alpha)^{\nu_1}$; and from (17) it is

[7] We shall say that the chain is *generated* by z_k.

seen that the nullspace of $(A - \alpha)^{\nu_1}$ has the basis $(x_j^i; j = 1, 2, \cdots, \nu_i, i = 1, 2, \cdots, p)$.

Since z_k belongs to the nullspace of $(A - \alpha)^{\nu_1}$, we may set

$$z_k = \sum_{i=1}^{p} \sum_{j=1}^{\nu_i} \zeta_{ij} x_j^i \tag{25}$$

and therefore by repeated application of (15) with $\lambda_i = \alpha$

$$(A - \alpha)^r z_k = \sum_{i,j} \zeta_{ij} x_{j-2}^i. \tag{26}$$

From this it follows that, in order that $(A - \alpha)^k z_k = 0$, only values of j which are less than or equal to k can actually occur in (25) and in order that $(A - \alpha)^{k-1} z_k \neq 0$ at least one ζ_{ik} must be different from 0; hence

$$\begin{aligned} z_k &= \sum_i (\zeta_{ik} x_k^i + \zeta_{i,k-1} x_{k-1}^i + \cdots) \\ z_{k-1} &= \sum_i (\zeta_{ik} x_{k-1}^i + \zeta_{i,k-1} x_{k-2}^i + \cdots) \\ &\cdots\cdots\cdots\cdots\cdots\cdots \\ z_1 &= \sum_i \zeta_{ik} x_1^i. \end{aligned} \tag{27}$$

Finally, if we impose the restriction that z_k does not belong to any chain pertaining to an exponent greater than k, it is necessary and sufficient that k be one of the numbers $\nu_1, \nu_2, \cdots, \nu_p$ and that no value of i corresponding to an exponent greater than k occur in (27).

3.09 The actual determination of the vectors x_j^i can be carried out by the processes of §3.02 and §3.04 or alternatively as follows. Suppose that the first s_1 of the exponents ν_i equal n_1, the next s_2 equal n_2, and so on, and finally the last s_q equal n_q. Let $\mathfrak{N}_1$ be the nullspace of $(A - \alpha)^{n_1}$ and $\mathfrak{N}_1'$ the nullspace of $(A - \alpha)^{n_1 - 1}$; then $\mathfrak{N}_1$ contains $\mathfrak{N}_1'$. If $\mathfrak{M}_1$ is a space complementary to $\mathfrak{N}_1'$ in $\mathfrak{N}_1$, then for any vector x in $\mathfrak{M}_1$ we have $(A - \alpha)^r x = 0$ only when $r \geq n_1$. Also, if $x_1, x_2, \cdots, x_{m_1}$ is a basis of $\mathfrak{M}_1$, the vectors

$$(A - \alpha)^r x_i \; (r = 0, 1, \cdots, n_1 - 1) \tag{28}$$

are linearly independent; for, if

$$\sum_{r=s}^{n_1-1} \sum_i \xi_{ir} (A - \alpha)^r x_i = 0,$$

some ξ_{ir} being different from 0, then multiplying by $(A - \alpha)^{n_1 - s - 1}$ we have

$$(A - \alpha)^{n_1 - 1} \sum_i \xi_{is} x_i = 0,$$

which is only possible if every $\xi_{is} = 0$ since $x_1, x_2, \cdots, x_{m_1}$ form a basis of $\mathfrak{M}_1$ and for no other vector of $\mathfrak{M}_1$ is $(A - \alpha)^{n_1 - 1} x = 0$. The space defined by (28) clearly lies in $\mathfrak{N}_1$; we shall denote it by $\mathfrak{L}_1$. If we set $\mathfrak{N}_1 = \mathfrak{N}_2 + \mathfrak{L}_1$ where $\mathfrak{N}_2$ is complementary to $\mathfrak{L}_1$ in $\mathfrak{N}_1$, then $\mathfrak{N}_2$ contains all vectors which are members of sets belonging to the exponents $n_2, n_3, \cdots$ but not lying in sets with the exponent n_1.

We now set $\mathfrak{N}_2 = \mathfrak{N}_2' + \mathfrak{M}_2$ where $\mathfrak{N}_2'$ is the subspace of vectors x in $\mathfrak{N}_2$ such that $(A - \alpha)^{n_2 - 1} x = 0$. As before the elements of $\mathfrak{M}_2$ generate sets with exponent n_2 but are not members of sets with higher exponents; and by a repetition of this process we can determine step by step the sets of invariant vectors corresponding to each exponent n_i.

CHAPTER IV

VECTOR POLYNOMIALS. SINGULAR MATRIC POLYNOMIALS

4.01 Vector polynomials. If a matric polynomial in λ is singular, the elements of its nullspace may depend on λ. We are therefore led to consider vectors whose coordinates are polynomials in a scalar variable λ; such a vector is called a *vector polynomial.* Any vector polynomial can be put in the form

$$z(\lambda) = z_0\lambda^m + z_1\lambda^{m-1} + \cdots + z_m$$

where $z_0, z_1, \cdots, z_m$ are vectors whose coordinates are independent of λ and, if $z_0 \neq 0$, m is called the degree of $z(\lambda)$. In a linear set with a basis composed of vector polynomials we are usually only concerned with those vectors that have integral coordinates when expressed in terms of the basis and, when this is so, we shall call the set an *integral set.* In a basis of an integral set the degree of an element of maximum degree will be called the degree of the basis.

In practice an integral set is often given in terms of a sequence of vectors which are not linearly independent and so do not form a basis. For the present therefore we shall say that the sequence of vector polynomials

$$z_1(\lambda), z_2(\lambda), \cdots, z_k(\lambda) \tag{1}$$

defines the integral set of all vectors of the form $\Sigma\zeta_i(\lambda)z_i(\lambda)$ where ζ's are scalar polynomials, and show later that this is really an integral set by finding for it an integral basis. The sequence (1) is said to have rank r if $| z_{i_1}z_{i_2} \cdots z_{i_s} |$ vanishes identically in λ for all choices of s z's when $s > r$ and is not identically 0 for some choice of the z's when $s = r$.

The theory of integral sets can be expressed entirely in terms of matric polynomials, but it will make matters somewhat clearer not to do so at first. By analogy with matrices we define an elementary transformation of a sequence of vector polynomials as follows. An elementary transformation of the sequence (1) is the operation of replacing it by a sequence $z_1', z_2', \cdots, z_k'$ where:

$$\text{Type I: } z_i' = z_i + \sum_{p\neq i} \zeta_p z_p,\ z_q' = z_q,\ (q \neq i),$$

$$\text{Type II: } z_i' = z_j,\ z_j' = z_i,\ z_q' = z_q\ (q \neq i, j),$$

$$\text{Type III: } z_p' = \rho_p z_p,\ (p = 1, 2, \cdots, k),$$

where the ζ's are scalar polynomials and the ρ's constants none of which is 0.

The rank of a sequence is not altered by an elementary transformation, and two sequences connected by an elementary transformation are equivalent in the sense that every vector polynomial belonging to the integral set defined by the one also belongs to the integral set defined by the other.

Two sequences which can be derived the one from the other by elementary transformations are said to be equivalent. The corresponding integral sets may also be said to be equivalent; and if only transformations with constant coefficients are used, the equivalence is said to be strict. Equivalence may also be defined as follows. If P is an elementary matrix which turns any vector of the integral linear set $(z_1, z_2, \cdots, z_k)$ into a vector of the same set, then it is easily shown that this set is equivalent to $(Pz_1, Pz_2, \cdots, Pz_k)$ and conversely; we also say that the linear set $(z_1, z_2, \cdots, z_k)$ is *invariant under* P although the individual elements of the basis are not necessarily unchanged. If the restriction that P leaves $(z_1, z_2, \cdots, z_k)$ invariant is not imposed, the two sets are said to be similar.

4.02 **The degree invariants.** We have seen in the previous section that the sequence in terms of which an integral set is defined may be transformed by elementary transformations without altering the integral set itself. We shall now show how we may choose a normalized basis and determine certain invariants connected with the set. Let the vectors (1), when written in full, be

$$z_i(\lambda) = \lambda^{m_i} z_{i0} + \lambda^{m_i - 1} z_{i1} + \cdots + z_{im_i}, \tag{2}$$

and suppose the notation so arranged that $m_1 \leq m_2 \leq \cdots \leq m_k$. Suppose further that the leading coefficients $z_{10}, z_{20}, \cdots, z_{s-1,0}$ are linearly independent but that

$$z_{s0} = \sum_1^{s-1} \eta_i z_{i0},$$

the η's being constants not all 0; then $m_s \geq m_i$ $(i = 1, 2, \cdots, s-1)$ and

$$z_s' = z_s - \sum_1^{s-1} \eta_i \lambda^{m_s - m_i} z_i$$

is either 0 or has a lower degree than z_s, and it may replace z_s in the sequence.

After a finite number of elementary transformations of this kind we arrive at a sequence equivalent to (1) which consists of a number p of vector polynomials $x_1, x_2, \cdots, x_p$ in which the leading coefficients are linearly independent followed by $k - p$ zero-vectors. Now if we form $| x_1 x_2 \cdots x_p |$ using the notation of (2) with x's in place of z's, the term of highest degree is $\lambda^{m_1 + \cdots + m_p} \cdot | x_{10} x_{20} \cdots x_{p0} |$, which is not 0 since the leading vectors $x_{10}, x_{20}, \cdots, x_{p0}$ are linearly independent. But the rank of a sequence is not changed by elementary transformations; hence $p = r$ and we have the following theorem.

Theorem 1. *If $z_1, z_2, \cdots, z_k$ is a sequence of vector polynomials of rank r, the set of vectors of the form $\sum_i \zeta_i(\lambda) z_i(\lambda)$, the ζ's being scalar polynomials, form an integral set with a basis of order r which may be so chosen that the leading coefficients of its constituent vectors are linearly independent.*

When a basis of an integral set satisfies the conditions of this theorem and its elements are arranged in order of ascending degree, we shall call it a *normal* basis.

Corollary. If $x_1, x_2, \cdots, x_r$ is a normal basis with the degrees $m_1 \leq m_2 \leq \cdots \leq m_r$, and if $\xi_1, \xi_2, \cdots, \xi_r$ are scalar polynomials, then the degree of the vector polynomial $x = \sum_1^s \xi_i x_i$ $(\xi_s \neq 0)$ is not less than m_s.

Theorem 2. *If $x_1, x_2, \cdots, x_r$ is a normal basis of an integral set and $m_1 \leq m_2 \leq \cdots \leq m_r$ the corresponding degrees, and if $y_1, y_2, \cdots, y_r$ is any other basis with the degrees $n_1 \leq n_2 \leq \cdots \leq n_r$, then*

$$m_1 \leq n_1, m_2 \leq n_2, \cdots, m_r \leq n_r.$$

Further, the exponents $m_1, m_2, \cdots, m_r$ are the same for all normal bases.

Let s be the first integer for which $n_s < m_s$ so that $n_i \leq n_s < m_s$ for $i \leq s$. Since $(x_1, x_2, \cdots, x_r)$ is a basis, we may set

$$y_i = \sum_p \xi_{ip}(\lambda)x_p(\lambda) \qquad (i = 1, 2, \cdots, s).$$

Here no value of p greater than $s - 1$ is admissible since the degree n_i of y_i is less than m_s. This would mean that the rank of $y_1, y_2, \cdots, y_s$ was less than s, which is impossible since they form part of a basis. Hence $m_s \leq n_s$ for all values of s.

If both bases are normal, it follows immediately that $m_i \leq n_i$ and also $n_i \leq m_i$, whence $m_i = n_i$, that is, the set of exponents $m_1, m_2, \cdots, m_r$ is the same for all normal bases. We shall call these exponents the *degree* invariants of the integral set.

4.03 **Elementary sets.** If $z_1(\lambda), z_2(\lambda), \cdots, z_r(\lambda)$ is a basis of an integral set, but not necessarily a normal basis, the r-vector $|z_1 z_2 \cdots z_r|$, which we call the determinant of the basis, is not identically 0 but may vanish for certain values of λ. If it vanishes for $\lambda = \lambda_1$, then $z_1(\lambda_1), z_2(\lambda_1), \cdots, z_r(\lambda_1)$ are linearly dependent, that is, there is a relation $\Sigma \zeta_i z_i(\lambda_1) = 0$; we may assume $\zeta_1 \neq 0$ without loss of generality. It follows that $\Sigma \zeta_i z_i(\lambda)$ has a factor of the form $(\lambda - \lambda_1)^\alpha$, $\alpha \geq 1$, and hence

$$z_1'(\lambda) = \frac{\Sigma \zeta_i z_i(\lambda)}{(\lambda - \lambda_1)^\alpha}$$

is integral; and, since $\zeta_1 \neq 0$, every element of $(z_1(\lambda), z_2(\lambda), \cdots, z_r(\lambda))$ is integrably expressible in terms of $(z_1'(\lambda), z_2(\lambda), \cdots, z_r(\lambda))$. Moreover, since

$$|z_1' z_2 \cdots z_r| = \frac{\zeta_1}{(\lambda - \lambda_1)^\alpha} |z_1 z_2 \cdots z_r|,$$

the determinant of the new basis is of lower degree than that of the old and so, if we continue this process, we shall arrive after a finite number of steps at a basis $(x_1(\lambda), x_2(\lambda), \cdots, x_r(\lambda))$ whose elements are linearly independent for all values of λ. A set which has a basis of this kind will be called an *elementary* integral set; and it is readily shown that every basis of an elementary integral set has the given property, namely, that its elements are linearly independent for every value of λ. These results are summarized as follows.

THEOREM 3. *Every integral set of order r is contained in an elementary set of the same order.*

We also have

THEOREM 4. *Let $x_1, x_2, \cdots, x_r$ be a basis of an elementary set. If $r < n$, there exists a complementary elementary basis $x_{r+1}, \cdots, x_n$ such that $|x_1x_2 \cdots x_n| \neq 0$ for any value of λ and this basis can be so chosen that its degree does not exceed that of $x_1, x_2, \cdots, x_r$.*

THEOREM 5. *If $x_1, x_2, \cdots, x_r$ is a basis of an elementary set, there exists an elementary matric polynomial X such that $x_i = Xe_i$ $(i = 1, 2, \cdots, r)$.*

For let y be a constant vector which for some value[1] of λ is not linearly dependent on $x_1, x_2, \cdots, x_r$ so that we do not have identically $y = \Sigma\eta_i x_i$ for any η's which are scalar polynomials. If for some value of λ, say λ_1, we have $y = \Sigma\xi_i x_i(\lambda_1)$, the ξ's being constants, then $y - \Sigma\xi_i x_i(\lambda)$ has the factor $\lambda - \lambda_1$ and, as in the proof of Theorem 3, we can modify y step by step till we arrive at a vector polynomial x_{r+1} such that $x_1, x_2, \cdots, x_r, x_{r+1}$ form an elementary basis. The degree at each step of this process does not exceed that of the original basis since only constant multipliers are used. This procedure may be continued till a basis of order n is reached, which proves Theorem 4.

The proof of Theorem 5 is immediate; in fact, using the basis derived in the proof of Theorem 4, $X = \sum_1^n x_i S e_i$ satisfies the required conditions and $|X| = |x_1x_2 \cdots x_n|$, which does not vanish for any value of λ.

As a converse to Theorem 5 we have that, *if X is an elementary matrix, then $x_i = Xe_i$ $(i = 1, 2, \cdots, r)$ is a basis of the elementary set $(x_1, x_2, \cdots, x_r)$.*

4.04 If $z_1, z_2, \cdots, z_k$ is a sequence of vector polynomials of rank r, we may always assume $k \leq n$ by merely increasing the order of the fundamental space, if necessary. Setting $z_i = \Sigma\zeta_{ji}e_j$, let us consider the matric polynomial

[1] If the question of degree is not important, any vector polynomial satisfying this condition may take the place of a constant vector.

$$Z = \left\| \begin{matrix} \zeta_{11} & \cdots & \zeta_{1k} & 0 & \cdots & 0 \\ \zeta_{21} & \cdots & \zeta_{2k} & 0 & \cdots & 0 \\ \cdots & \cdots & \cdots & \cdots & \cdots & \cdots \\ \cdots & \cdots & \cdots & \cdots & \cdots & \cdots \\ \zeta_{n1} & \cdots & \zeta_{nk} & 0 & \cdots & 0 \end{matrix} \right\| = \sum_1^k z_i S e_i.$$

The elementary transformations used in §4.02 in finding a basis of the integral set correspond when applied to Z to a combination of elementary transformations, as defined as §3.01, and because these transformations involve columns only, they correspond to multiplying Z on the right by an elementary polynomial Q_1. Similarly, if

$$Z = \Sigma e_i S y_i,\ y_i = \Sigma \zeta_{ij} e_j,$$

the process of finding a basis for $y_1, y_2, \cdots, y_n$, whose rank is r, corresponds to multiplying Z on the left by an elementary polynomial P_1.

We shall now suppose that $k = r$ so that $Q_1 = 1$; then $P_1 Z$ has the form

$$Z_1 = P_1 Z = \left\| \begin{matrix} \omega_{11} & \cdots & \omega_{1r} & 0 & \cdots & 0 \\ \omega_{21} & \cdots & \omega_{2r} & 0 & \cdots & 0 \\ \cdot & \cdots & \cdot & \cdot & \cdots & \cdot \\ \omega_{r1} & \cdots & \omega_{rr} & 0 & \cdots & 0 \\ 0 & \cdots & 0 & 0 & \cdots & 0 \\ \cdot & \cdots & \cdot & \cdot & \cdots & \cdot \\ 0 & \cdots & 0 & 0 & \cdots & 0 \end{matrix} \right\|$$

We now bring Z_1 to the normal form of §3.02, say

$$PZ_1Q = PP_1ZQ = \left\| \begin{matrix} \zeta_1 & & & & & & & \\ & \zeta_2 & & & & & & \\ & & \cdot & & & & & \\ & & & \cdot & & & & \\ & & & & \zeta_r & & & \\ & & & & & 0 & & \\ & & & & & & \cdot & \\ & & & & & & & \cdot \\ & & & & & & & & 0 \end{matrix} \right\|$$

where $\zeta_1, \zeta_2, \cdots, \zeta_r$ are the invariant factors of Z (or Z_1) and in doing so only the first r rows and columns are involved so that

$$Qe_i = e_i,\ (i = r + 1, r + 2, \cdots, n). \tag{3}$$

Therefore, if $x_i = (PP_1)^{-1}e_i$, we have successively

$$PP_1ZQe_i = \zeta_ie_i,\ ZQe_i = \zeta_i(PP_1)^{-1}e_i = \zeta_ix_i,\quad (i = 1, 2, \cdots, r)$$

and, if $Q = \| q_{ij} \|$,

$$Qe_i = q_{1i}e_1 + q_{2i}e_2 + \cdots + q_{ri}e_r,\quad (i = 1, 2, \cdots, r)$$

and hence

$$(4)\qquad \zeta_ix_i = ZQe_i = q_{1i}z_1 + q_{2i}z_2 + \cdots + q_{ri}z_r,\ (i = 1, 2, \cdots, r).$$

But from (3) and the fact that $| Q |$ is a constant different from 0, it follows that the determinant of the coefficients in (4) is also a constant different from 0, and hence these equations can be solved for the z's in terms of the x's giving, say

$$z_i = \Sigma b_{ji}\zeta_jx_j\ (i = 1, 2, \cdots, r)$$

where the b's are scalar polynomials.

Returning now to the case $k \geq r$, we see that, since we can pass to the case $k = r$ by elementary transformations, the ζ's are still the invariant factors of $Z = \sum_1^k z_iSe_i$. They are therefore also invariants of the integral set independently of the basis chosen to represent it, and so we shall call them the *invariant factors* of the set.

We can now state the following theorem.

THEOREM 6. *If $\zeta_1, \zeta_2, \cdots, \zeta_r$ are the invariant factors of an integral set of vector polynomials, we can find a basis of the form*

$$\zeta_1x_1, \zeta_2x_2, \cdots, \zeta_rx_r$$

where $x_1, x_2, \cdots, x_r$ define an elementary set.

4.05 Linear elementary bases. We shall derive in this section a canonical form for a basis of an elementary linear set. If

$$(5)\qquad z_1, z_2, \cdots, z_r,\ z_i = x_i\lambda + y_i,$$

is a basis of an elementary linear set, it is convenient, though not necessary, to associate with it the matrix

$$(6)\qquad A\lambda - B = \sum_1^r g_iSz_i = \sum_1^r g_iS(x_i\lambda + y_i)$$

where $g_1, g_2, \cdots, g_r$ is a sequence of linearly independent constant vectors. When this is done, it should be noted that multiplying $A\lambda - B$ on the right by an elementary matrix P corresponds to replacing (5) by the similar sequence $P'z_1, \cdots, P'z_r$. Multiplying on the left by P has no immediate interpretation in terms of the sequence except when

$$Pg_i = \sum_{j=1}^r p_{ij}g_j\quad (i = 1, 2, \cdots, r)$$

in which case we can write

$$P(A\lambda - B) = \sum_{i=1}^{r} g_i S \sum_{j=1}^{r} p_{ji} z_j = \sum_{i=1}^{r} g_i S z'_i$$

and the set $(z'_1, z'_2, \cdots, z'_r)$ is equivalent to (5); when P is constant, the equivalence is strict.

Instead of restricting ourselves to the matrix (6), we shall only assume to begin with that $A\lambda - B$ is a linear matric polynomial of rank $r < n$. The nullspace $\mathfrak{N}$ of $A\lambda - B$ is then an elementary integral set, a normalized basis of which we shall take to be

(7) $$a_i(\lambda) = a_{i0}\lambda^{m_i} + a_{i1}\lambda^{m_i - 1} + \cdots + a_{im_i} \quad (i = 1, 2, \cdots, n - r).$$

From $(A\lambda - B)a_i = 0$ we have

$$Aa_{i0} = 0, \; Aa_{i1} = Ba_{i0}, \; \cdots, \; 0 = Ba_{im_i},$$

or, if we set $a_{it} = 0$ for $t < 0$ or $t > m_i$,

(8) $$Aa_{it} = Ba_{i,\,t-1} \quad (t = 0, 1, 2, \cdots, m_i + 1).$$

We shall now show that the vectors a_{ij} $(i = 1, 2, \cdots, n - r; j = 0, 1, \cdots, m_i)$ are linearly independent. Assume that a_{ij} are linearly independent for $(i = 1, 2, \cdots, p - 1; j = 0, 1, \cdots, m_i)$ and $(i = p; j = 0, 1, \cdots, q - 1)$ but that

(9) $$a_{pq} + \sum_{j=0}^{q-1} \alpha_{pj} a_{pj} + \sum_{i=1}^{p-1} \sum_{j=0}^{m_i} \alpha_{ij} a_{ij} = 0.$$

Let s' be the greatest value of j for which some $\alpha_{ij} \neq 0$ and let s be the greater of s' and q. If we set

(10) $$c_t = a_{p,q-s+t} + \sum_{j=0}^{q-1} \alpha_{pj} a_{p,\,j-s+t} + \sum_{i=1}^{p-1} \sum_{j=0}^{m_i} \alpha_{ij} a_{i,\,j-s+t},$$

then $c_{-1} = 0$, $c_s = 0$ and

$$c_{s-1} = a_{p,\,q-1} + \sum_{j=1}^{q-1} \alpha_{pj} a_{p,\,j-1} + \sum_{i=1}^{p-1} \sum_{j=1}^{m_i} \alpha_{ij} a_{i,\,j-1},$$

which is not 0 by hypothesis, except perhaps when $q = 0$ and every α_{ij} $(j \neq 0)$ in (9) is 0, which, however, is not possible since by Theorem 1 the leading coefficients a_{i0} in (7) are linearly independent. Also from (8) it follows that $Ac_t = Bc_{t-1}$, and hence

(11) $$c(\lambda) = c_0\lambda^{s-1} + c_1\lambda^{s-2} + \cdots + c_{s-1}$$

is a null-vector of $A\lambda - B$ of degree less than m_p. But every such integral null-vector is linearly dependent on $a_1, a_2, \cdots, a_{p-1}$ with integral coefficients, say

$$c(\lambda) = \sum_1^{p-1} \gamma_i(\lambda) a_i(\lambda); \tag{12}$$

and this gives

$$c_{s-1} = c(0) = \sum_1^{p-1} \gamma_i(0) a_i(0) = \sum_1^{p-1} \gamma_i(0) a_{im_i},$$

which is impossible since c_{s-1} is obtained from (9) by lowering the second subscript in each term and no such subscript greater than m_i can occur in any a_{ij}. Hence the a_{ij} are linearly independent.

In order to simplify the notation we shall now set

$$a_{ij} = Qe_j^i \ (i = 1, 2, \cdots, n - r; j = 0, 1, \cdots, m_i) \tag{13}$$

where Q is a constant non-singular matrix and e_j^i are fundamental units rearranged by setting, say, $e_k = e_j^i$ when $k = \sum_{q=1}^{i-1} (m_q + 1) + j + 1$; as before $e_j^i = 0$ for $j < 0$ and $j > m_i$. We shall denote the space defined by the e_j^i by $\mathfrak{M}_1$ and the complementary space by $\mathfrak{M}_2$; since the bases of $\mathfrak{M}_1$ and $\mathfrak{M}_2$ can be chosen as sequences of fundamental units, they are reciprocal as well as complementary.

We return to the particular case in which $A\lambda - B$ is given by (6). Corresponding to (12) we define a new set of vectors w by

$$w_i = Q'z_i \quad (i = 1, 2, \cdots, r) \tag{14}$$

and when this is done a normal basis of the nullspace $\mathfrak{N}_1$ of $(A\lambda - B)Q = \Sigma g_i S w_i$ is given by

$$b_i = e_0^i \lambda^{m_i} + e_1^i \lambda^{m_i - 1} + \cdots + e_{m_i}^i \ (i = 1, 2, \cdots, n - r). \tag{15}$$

We have seen in §1.10 that $b_1, b_2, \cdots, b_{n-r}$ is the space reciprocal to $w_1, w_2, \cdots, w_r$. Now in $\mathfrak{M}_1$ the $\sum_1^{n-r} m_i$ vectors

$$f_j^i = e_{j-1}^i - \lambda e_j^i \quad (i = 1, 2, \cdots, n - r; j = 1, 2, \cdots, m_i) \tag{16}$$

are linearly independent; and they form the set reciprocal to (14) in $\mathfrak{M}_1$ since $Sf_j^i b_p = 0$ for all i, j, p and the sum of the orders of the two sets is $\Sigma m_i + (n - r)$ which is the order of $\mathfrak{M}_1$. Hence the total set $(w_1, w_2, \cdots, w_r)$ reciprocal to $(b_1, b_2, \cdots, b_{n-r})$ is composed of (15) together with $\mathfrak{M}_2$. We shall call this form of basis a *canonical* basis of the set (13). We can now state the following theorems.

THEOREM 7. *A linear elementary set of order r has a basis of the form*

(17) g_j^0 $(j = 1, 2, \cdots, m)$, $g_j^i = a_{i,\,j-1} + \lambda a_{ij}$ $(i = 1, 2, \cdots, \nu; j = 1, 2, \cdots, m_i)$

where the constant vectors g_j^0, a_{ij} are linearly independent for all j and i and the integers m_i are those degree invariants of the reciprocal set that are not 0.

We shall call each set g_j^i $(j = 1, 2, \cdots, m_i)$ a *chain* of *index* m_i, and define the integers $m_1, m_2, \cdots, m_\nu$ as the *Kronecker invariants* of the set. A basis of the form (17) will be called canonical.

THEOREM 8. *Two linear elementary sets are similar if, and only if, they have the same Kronecker invariants and the same order.*

It should be noted that, if r is the order of the set, then

$$m + \sum_1^\nu m_i = r, \; m + \sum_1^\nu (m_i + 1) \le n, \; \nu \le n - r. \tag{18}$$

If $r = n$, all the Kronecker invariants are 0 and there are no chains in the basis.

If $z_1, z_2, \cdots, z_r$ is a normal basis of an elementary linear set, the first m being constant and the rest linear in λ, and g_j^0, g_j^i is a canonical basis, the notation being that of Theorem 7, then clearly the set g_j^0 $(j = 1, 2, \cdots, m)$ is strictly equivalent to $(z_1, z_2, \cdots, z_m)$ and the remaining vectors have the form

$$g_j^i = w_j^i + u_j^i + \lambda v_j^i \tag{19}$$

where u_j^i and v_j^i belong to $(z_1, z_2, \cdots, z_m)$ and the w_j^i are constant linear combinations of $z_{m+1}, \cdots, z_r$. Since a canonical basis is also normal,

(20) g_j^0 $(j = 1, 2, \cdots, m)$, w_j^i $(i = 1, 2, \cdots, \nu; j = 1, 2, \cdots, m_i)$

is a normal basis strictly equivalent to $(z_1, z_2, \cdots, z_r)$. Now (19) may be written

$$w_j^i = g_j^i - u_j^i - \lambda v_j^i = a_{i,\,j-1} - u_j^i + \lambda(a_{ij} - v_j^i) = z_j^i + b_j^i \tag{21}$$

where $b_j^i = v_{j-1}^i - u_j^i$ is a constant vector of the linear set $(g_1^0, g_2^0, \cdots, g_m^0)$ and

$$z_j^i = a_{i,\,j-1} - v_{j-1}^i + \lambda(a_{ij} - v_j^i). \tag{22}$$

Here (22) together with the g_j^0 form a canonical basis which from (21) is strictly equivalent to (20) and therefore to $(z_1, z_2, \cdots, z_r)$. We therefore have the following theorem.

THEOREM 9. *Every normal basis of a linear elementary set is strictly equivalent to some canonical basis.*

4.06 Singular linear polynomials. Let $A\lambda + B$ be a matric polynomial of rank $r < n$. Its left and right grounds are linear integral sets of rank r, and by Theorems 3 and 7 we can find canonical bases in terms of which the vectors of the two grounds can be integrally expressed, say

$$z_1, z_2, \cdots, z_r \quad \text{and} \quad w_1, w_2, \cdots, w_r, \tag{23}$$

respectively, where the first α z's and β w's are constant and the rest linear in λ. When $A\lambda + B$ is expressed in terms of these bases, then, remembering that no second degree term can appear, we see that it has the form

$$(24)\quad \sum_{i=1}^{\alpha}\sum_{j=1}^{\beta}(h_{ij}\lambda + k_{ij})z_iSw_j + \sum_{i=1}^{r-\alpha}\sum_{j=1}^{\beta}k_{\alpha+i,j}z_{\alpha+i}Sw_j + \sum_{i=1}^{\alpha}\sum_{j=1}^{r-\beta}k_{i,\beta+j}z_iSw_{\beta+j}.$$

The row vectors

$$(25')\qquad p_i = \sum_{j=1}^{\beta}(h_{ij}\lambda + k_{ij})w_j + \sum_{j=1}^{r-\beta}k_{i,\beta+j}w_{\beta+i} \qquad (i = 1, 2, \cdots, \alpha)$$

$$(25'')\qquad p_{\alpha+i} = \sum_{j=1}^{\beta}k_{\alpha+i,j}w_j \qquad (i = 1, 2, \cdots, r - \alpha)$$

form a set of r linearly independent vectors and, since the set (25″) depends only on β w's, we must have $r - \alpha \leq \beta$. Setting $\gamma = \alpha + \beta - r$ we may replace $w_{\gamma+1}, \cdots, w_\beta$ by $p_{\alpha+1}, \cdots, p_r$ in (23) without destroying the canonical form of the basis. A similar change can be made independently in the z basis by replacing $z_{\gamma+j}$ by $\sum_{1}^{\alpha} k_{i,\beta+\gamma}z_{\beta+i}$ $(j = 1, 2, \cdots, r - \beta = \alpha - \gamma)$.

When we assume that these changes have been made to begin with, we may take in place of (24)

$$(26)\qquad A\lambda + B = \sum_{i=1}^{\alpha}\sum_{j=1}^{\beta}(h_{ij}\lambda + k_{ij})z_iSw_j + \sum_{i=1}^{r-\alpha}z_{\alpha+i}Sw_{\gamma+i} + \sum_{j=1}^{r-\beta}z_{\gamma+j}Sw_{\beta+j}.$$

Figure 1 shows schematically the effect of this change of basis. To begin with the coefficients in (24) may be arranged in a square array AR of side r; the

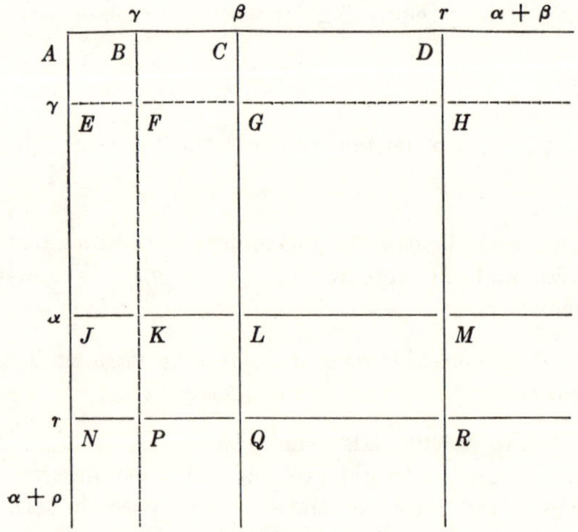

FIG. 1

first double sum corresponds to the rectangle AL, the second to JQ and the third to CM, and the rectangle LR contains only zeros. After the transformation which leads to (26), the only change in the scheme is that in JQ the part JP is now zero and the square KQ has 1 in the main diagonal and zeros elsewhere, and CM also takes a similar form.

If we set

$$z'_{\alpha+i} = z_{\alpha+i} + \sum_{j=1}^{\alpha} (h_{j,\gamma+i}\lambda + k_{j,\gamma+i})z_j \qquad (i = 1, 2, \cdots, r - \alpha)$$

$$w'_{\beta+i} = w_{\beta+i} + \sum_{j=1}^{\gamma} (h_{\gamma+i,j}\lambda + k_{\gamma+i,j})w_j \quad (i = 1, 2, \cdots, r - \beta)$$

then $z_1, \cdots, z, z'_{\alpha+1}, \cdots, z'_r$ and $w_1, \cdots, w_\beta, w'_{\beta+1}, \cdots, w'_r$ are still elementary bases of the right and left grounds, and in terms of them (26) becomes

$$\sum_{i,j=1}^{\gamma} (h_{ij}\lambda + k_{ij})z_iSw_j + \sum_{i=1}^{r-\alpha} z'_{\alpha+i}Sw_{\gamma+i} + \sum_{i=1}^{r-\beta} z_{\gamma+i}Sw'_{\beta+i}.$$

The number of terms in these summations after summing for j is $\gamma + (r - \alpha) + (r - \beta) = r$. Hence the rank of the square array $h_{ij}\lambda + k_{ij}$ $(i, j = 1, 2, \cdots, \gamma)$ is γ and by a change of variable of the form $\lambda - \lambda_1 = \lambda'$, if necessary, we can secure that the array k_{ij} is also of rank γ.

The transformation just employed disturbs the canonical form of the basis and we have now to devise a different transformation which will avoid this. Let us set in place of $w_1, w_2, \cdots, w_\gamma$

$$w'_j = w_j - \sum_{t=\gamma+1}^{\beta} p_{j,t-\gamma}w_t \qquad (j = 1, 2, \cdots, \gamma)$$

where the p's are constants to be determined later, and for brevity set also

$$h_j = \sum_{i=1}^{\gamma} h_{ij}z_i, \; k_j = \sum_{i=1}^{\gamma} k_{ij}z_i \qquad (j = 1, 2, \cdots, \beta);$$

since the rank of k_{ij} $(i = 1, 2, \cdots, \gamma)$ is γ, the vectors $k_1, k_2, \cdots, k_\gamma$ form a basis of $(z_1, z_2, \cdots, z_\gamma)$. After this change of basis the part of the first double sum (cf. (26)) which corresponds to $i = 1, 2, \cdots, \gamma; j = \gamma + 1, \cdots, \beta$ is

$$\text{(27)} \qquad \sum_{j=\gamma+1}^{\beta} \left[h_j\lambda + k_j + \sum_{t=1}^{\gamma} p_{t,j-\gamma}(h_t\lambda + k_t) \right] Sw_j.$$

Consider now a single chain of z's of index s which by a suitable change of notation we may suppose to be $z_{\alpha+1}, z_{\alpha+2}, \cdots, z_{\alpha+s}$; we shall seek to determine the p's so that the corresponding part of (27) shall become

$$\text{(28)} \qquad \sum_{j=\gamma+1}^{\gamma+s} (g_{j-1-\gamma} + \lambda g_{j-\gamma})Sw_j,$$

the g's being vectors in the space $(z_1, z_2, \cdots, z_\gamma)$. Equating corresponding terms in (27) and (28) we have

$$g_0 = k_{\gamma+1} + \sum_t p_{t1}k_t,\ g_1 = h_{\gamma+1} + \sum_t p_{t1}h_t = k_{\gamma+2} + \sum_t p_{t2}k_t$$

$$\cdots\cdots\cdots\cdots\cdots\cdots\cdots\cdots$$

$$g_{s-1} = h_{\gamma+s-1} + \sum_t p_{t,s-1}h_t = k_{\gamma+s} + \sum_t p_{ts}k_t,\ g_s = h_{\gamma+s} + \sum_t p_{ts}h_t.$$

Choosing p_{t1} $(t = 1, 2, \cdots, \gamma)$ arbitrarily we define g_0 by the first equation; then the second defines p_{t2} since the vector $h_{\gamma+1} + \sum_t p_{t1}h_t - k_{\gamma+2}$ can be expressed uniquely in terms of the basis $(k_1, k_2, \cdots, k_\gamma)$; and the remaining p's are similarly determined in succession, while the last equation defines g_s.

If we now in our basis put in place of $z_{\alpha+i}$

$$z'_{\alpha+i} = z_{\alpha+i} + g_{i-1} + \lambda g_i \qquad (i = 1, 2, \cdots, s)$$

and combine the corresponding part of (27) with $\sum_{i=1}^{s} z_{\alpha+i}Sw_{\gamma+i}$, the two together give $\sum_{i=1}^{s} z'_{\alpha+i}Sw_{\gamma+i}$ and the new basis is still canonical. We then treat all the z chains in the same way and have finally in place of (26)

$$\sum_{i,j=1}^{\gamma} (h_{ij}\lambda + k_{ij})z_iSw'_j + \sum_{j=\gamma+1}^{\alpha} \sum_{j=1}^{\beta} (h_{ij}\lambda + k_{ij})z_iSw'_j + \sum_{i=1}^{r-\alpha} z'_{\alpha+i}Sw_{\gamma+i}$$
$$+ \sum_{j=1}^{r-\beta} z_{\gamma+j}Sw_{\beta+j}.$$

The changes in the bases used above have replaced the coordinates h_{ij}, k_{ij} by 0 for the range $i = 1, 2, \cdots, \gamma; j = \gamma + 1, \cdots, \beta$ and have left them wholly unaltered for $i = \gamma + 1, \cdots, \alpha; j = 1, 2, \cdots, \gamma$. We can therefore interchange the rôles of the z's and w's and by modifying now the w-chains we can make these coordinates zero for the second range of subscripts without altering the zeros already obtained for the first range. Hence it is possible by a suitable choice of the original canonical bases to assume that (26) is replaced by

$$A\lambda + B = \sum_{i,j=1}^{\gamma} (h_{ij}\lambda + k_{ij})z_iSw_j + \sum_{i=\gamma+1}^{\alpha} \sum_{j=\gamma+1}^{\beta} (h_{ij}\lambda + k_{ij})z_iSw_j$$
$$+ \sum_{i=1}^{r-\alpha} z_{\alpha+i}Sw_{\gamma+i} + \sum_{j=1}^{r-\beta} z_{\gamma+j}Sw_{\beta+j}. \qquad (29)$$

Here the first summation can be reduced to a canonical form without affecting the rest of the expression; we therefore neglect it in the meantime and deal only with the remaining terms. This is equivalent to putting $\gamma = 0$, and in making this change it is convenient to alter the notation so as to indicate the chains in the bases.

As in §4.05 let the chains of the z and w bases be

$$f^i_{j-1} + \lambda f^i_j \qquad (j = 1, 2, \cdots, s_i;\ i = 1, 2, \cdots, \nu_1)$$

and

$$g^p_{q-1} + \lambda g^p_q \qquad (q = 1, 2, \cdots, t_p;\ p = 1, 2, \cdots, \nu_2)$$

respectively, and denote the constant vectors of the respective bases by z^p_q and w^i_j where i, j, p, q take the values indicated above since, when $\gamma = 0$, we have $\sum_1^{\nu_1} s_j = r - \alpha = \beta$, $\sum_1^{\nu_2} t_q = r - \beta = \alpha$. We have then to determine a canonical form for the matrix

$$(30) \quad \sum_{i,j,p,q} (h^{p\,i}_{q\,j}\lambda + k^{p\,i}_{q\,j})z^p_q S w^i_j + \sum_{i,j} (f^i_{j-1} + \lambda f^i_j) S w^i_j + \sum_{p,q} z^p_q S(g^p_{q-1} + \lambda g^p_q),$$

and in doing so we shall show that the first summation can be eliminated by a proper choice of the bases of the chains.

It will simplify the notation if we consider first only two chains, one of index s in the z-basis and the other of index t in the w-basis and, omitting the superscripts, choose the notation so that these chains are $f_0 + \lambda f_1, \cdots, f_{s-1} + \lambda f_s$ and $g_0 + \lambda g_1, \cdots, g_{t-1} + \lambda g_t$. We now modify these by adding $a_{j-1} + \lambda a_j$ to $f_{j-1} + \lambda f_j$ and $b_{i-1} + \lambda b_i$ to $g_{i-1} + \lambda g_i$ choosing

$$a_j = \sum_{i=1}^{t} \alpha_{ij} z_i \qquad (j = 0, 1, \cdots, s), \qquad b_i = \sum_{j=1}^{s} \beta_{ij} w_j \qquad (i = 0, 1, \cdots, t)$$

in such a way as to eliminate the corresponding terms in the first summation of (30). To do this we must choose the α's and β's so that

$$(31) \quad h_{ij} = \alpha_{ij} + \beta_{ij}, \qquad k_{ij} = \alpha_{i,j-1} + \beta_{i-1,j} \qquad (i = 1, 2, \cdots, t;\ j = 1, 2, \cdots, s).$$

For $j > 1$ this gives $\alpha_{i,j-1} = h_{i,j-1} - \beta_{i,j-1}$ and hence if $l_{ij} = k_{ij} - h_{i,j-1}$ we may write

$$\left.\begin{aligned} &(32') \quad k_{i1} = l_{i1} = \alpha_{i0} + \beta_{i-1,1} \\ &(32'') \quad l_{ij} = \beta_{i-1,j} - \beta_{i,j-1} \end{aligned}\right\} \qquad (i = 1, 2, \cdots, t;\ j = 2, 3, \cdots, s).$$

If we give α_{i0} $(i = 1, 2, \cdots, t)$ arbitrary values, (32′) defines β_{i1} for $i = 0, 1, \cdots, t - 1$ and leaves β_{t1} arbitrary; then $j = 2$ in (32″) gives β_{i2} for $i = 0, 1, \cdots, t - 1$ and leaves β_{t2} arbitrary, and so on: and when the β's are found in this way, certain of them being arbitrary, the first equation of (31) gives the remaining α's.

Combining every z chain in this way with each w chain in turn, we finally eliminate all the terms in the quadruple sum in (30), and (29) may therefore, by a proper choice of the two bases, be replaced by

(33)
$$A\lambda + B = \sum_{i,j=1}^{\gamma} (h_{ij}\lambda + k_{ij})z_i S w_j + \sum_{i,j} (f^i_{j-1} + \lambda f^i_j) S w^i_j + \sum_{p,q} z^p_q S(g^p_{q-1} + \lambda g^p_q)$$

where no two of the linear sets

(34′) $$(z_1, z_2, \cdots, z_\gamma), \qquad (f^i_0, f^i_1, \cdots, f^i_{s_i}, i = 1, 2, \cdots, \nu_1), \qquad (z^p_1, z^p_2, \cdots, z^p_{t_p}, p = 1, 2, \cdots, \nu_2)$$

have any vector in common, and also no two of

(34″) $$(w_1, w_2, \cdots, w_\gamma), \qquad (w^i_1, w^i_2, \cdots, w^i_{s_i}, i = 1, 2, \cdots, \nu_1), \qquad (g^p_0, g^p_1, \cdots, g^p_{t_p}, p = 1, 2, \cdots, \nu_2)$$

have any vector in common.

We shall now for the moment suppose that the order n of the fundamental space is taken so large that we can introduce vectors z^p_0 $(p = 1, 2, \cdots, \nu_2)$ into the third set in (34′) without causing the three spaces to overlap, and also w^i_0 $(i = 1, 2, \cdots, \nu_1)$ into the second set of (34″). As a matter of convenience we can then find two constant non-singular matrices P, Q such that

$$Pz_i = e_i = Q'w_i, \qquad Pf^i_j = e^i_j = Q'w^i_j, \qquad Pz^p_q = e^{\nu_1+p}_q = Q'g^p_q$$

where the range of the affices is as in (34) and where

$$e^i_j = e_k, \quad k = \gamma + \sum_{\alpha=1}^{i-1} (s_\alpha + 1) + j + 1, \quad e^{\nu_1+p}_q = e_k,$$

$$k = \gamma + \sum_{i=1}^{\nu_1} (s_i + 1) + \sum_{\alpha=1}^{p-1} (t_\alpha + 1) + q + 1$$

and, when this is done,

(35) $$P(A\lambda + B)Q = \sum_{i,j=1}^{\gamma} (h_{ij}\lambda + k_{ij})e_i S e_j + \sum_{i,j} (e^i_{j-1} + \lambda e^i_j) S e^i_j + \sum_{p,q} e^{\nu_1+p}_q S(e^{\nu_1+p}_{q-1} + \lambda e^{\nu_1+p}_q).$$

This matrix is composed of a number of blocks of terms arranged along the main diagonal, the remaining coordinates being 0. It must be carefully observed however that, owing to the introduction of the vectors z^p_0, w^i_0 and to the fact that a chain of index s depends on $s + 1$ constant vectors, the total number of rows and columns employed is greater than the rank r by the total number of chains in the left and right grounds.

The first summation in (35) gives a block $\| h_{ij}\lambda + k_{ij} \|$ of γ rows and columns. Each chain in the second and third summation gives a block of the respective forms

$$
(36) \qquad
\begin{matrix}
0 & 1 & 0 & \cdots & 0 & 0 \\
0 & \lambda & 1 & \cdots & 0 & 0 \\
0 & 0 & \lambda & \cdots & 0 & 0 \\
\cdots & \cdots & \cdots & \cdots & \cdots & \cdots \\
\cdots & \cdots & \cdots & \cdots & \cdots & \cdots \\
0 & 0 & 0 & \cdots & \lambda & 1 \\
0 & 0 & 0 & \cdots & 0 & \lambda
\end{matrix}
\qquad\qquad
\begin{matrix}
0 & 0 & 0 & \cdots & 0 & 0 \\
1 & \lambda & 0 & \cdots & 0 & 0 \\
0 & 1 & \lambda & \cdots & 0 & 0 \\
\cdots & \cdots & \cdots & \cdots & \cdots & \cdots \\
\cdots & \cdots & \cdots & \cdots & \cdots & \cdots \\
0 & 0 & 0 & \cdots & \lambda & 0 \\
0 & 0 & 0 & \cdots & 1 & \lambda.
\end{matrix}
$$

If we take $A\lambda + B\mu$ in place of $A\lambda + B$ and calculate the invariant factors, these forms show that we obtain the invariant factors of $\sum_i (h_{ij}\lambda + k_{ij}\mu)e_iSe_j$ together with a number of 1's from the blocks of type (36), the number contributed by each being one less than the number of rows it contains, that is, the index of the corresponding chain. This gives the following theorem.

THEOREM 10. *Two matric polynomials $A\lambda + B\mu$ and $C\lambda + D\mu$ are strictly equivalent if, and only if, they have the same invariant factors and their respective right and left grounds have the same Kronecker invariants.*

That these conditions are necessary is obvious; that they are sufficient follows readily from the form (33) derived above. In the first place, since the Kronecker invariants are the same for both, the second and third summations in (33) have the same form for both and are therefore strictly equivalent by means of transformations which do not change the terms in the first summation. Secondly, the first summation in both yields the same invariant factors since the number of 1's due to the remaining terms depends only on the number of chains, which is the same for both; hence these summations are strictly equivalent and, because of the linear independence of the constant vectors involved, the equivalence is obtainable by transformations which do not affect the remaining terms.

When the first summation in (35) is in canonical form, we shall say that $A\lambda + B$ is in its canonical form. This is however not altogether satisfactory since the space necessary for this form may be of greater order than n. If ν is the greater of ν_1 and ν_2, (33) shows that the minimum order of the enveloping space is $\gamma + \Sigma s_i + \Sigma t_p + \nu$. A canonical form for this number of dimensions can be obtained as follows. Pair the blocks of the first and second types of (36) till all of one type are used up, taking the order of the constituents in, say, the order of (36); then in the composite block formed from such a pair discard the first column and also the row which contains the first row of the second block. This gives a canonical form for such a pair, namely,

$$
(37)\qquad
\begin{array}{cccccccccc}
1 & 0 & \cdots & 0 & 0 & 0 & 0 & \cdots & 0 & 0 \\
\lambda & 1 & \cdots & 0 & 0 & 0 & 0 & \cdots & 0 & 0 \\
0 & \lambda & \cdots & 0 & 0 & 0 & 0 & \cdots & 0 & 0 \\
. & . & \cdots & . & . & . & . & \cdots & . & . \\
0 & 0 & \cdots & 1 & 0 & 0 & 0 & \cdots & 0 & 0 \\
0 & 0 & \cdots & \lambda & 0 & 0 & 0 & \cdots & 0 & 0 \\
0 & 0 & \cdots & 0 & 1 & \lambda & 0 & \cdots & 0 & 0 \\
0 & 0 & \cdots & 0 & 0 & 1 & \lambda & \cdots & 0 & 0 \\
. & . & \cdots & . & . & . & . & \cdots & . & . \\
0 & 0 & \cdots & 0 & 0 & 0 & 0 & \cdots & \lambda & 0 \\
0 & 0 & \cdots & 0 & 0 & 0 & 0 & \cdots & 1 & \lambda.
\end{array}
$$

If the number of chains in the left and right grounds is not the same, there will of course be blocks of one of the types (36) left unpaired.

CHAPTER V

COMPOUND MATRICES

5.01 In chapter I it was found necessary to consider the adjoint of A which is a matrix whose coordinates are the first minors of $|A|$. We shall now consider a more general class of matrices, called compound matrices, whose coordinates are minors of $|A|$ of the rth order; before doing so, however, it is convenient to extend the definition of Sxy to apply to vectors of higher grade.

5.02 **The scalar product.** Let $x_i = \Sigma\xi_{ij}e_j$, $y_i = \Sigma\eta_{ij}e_j$ $(i = 1, 2, \cdots)$ be arbitrary vectors, then, by equation (37) §1.11 we have

$$(1)\qquad |x_1x_2 \cdots x_r| = \sum_{(i)}^{*} |\xi_{1i_1}\xi_{2i_2} \cdots \xi_{ri_r}|\,|e_{i_1}e_{i_2} \cdots e_{i_r}|,$$

and hence it is natural to extend the notion of the scalar product by setting

$$(2)\qquad S|x_1x_2 \cdots x_r|\,|y_1y_2 \cdots y_r| = \sum_{(i)}^{*} |\xi_{1i_1}\xi_{2i_2} \cdots \xi_{ri_r}|\,|\eta_{1i_1}\eta_{2i_2} \cdots \eta_{ri_r}|.$$

We then have the following lemma which becomes the ordinary rule for multiplying together two determinants when $r = n$.

LEMMA 1.

$$(3)\qquad S|x_1x_2 \cdots x_r|\,|y_1y_2 \cdots y_r| = |Sx_iy_j|.$$

For $S|x_1x_2 \cdots x_r|\,|e_{i_1}e_{i_2} \cdots e_{i_r}| = |\xi_{1i_1}\xi_{2i_2} \cdots \xi_{ri_r}|$, hence

$$S|x_1x_2 \cdots x_r|\,|y_1e_{i_2} \cdots e_{i_r}| = \sum_{i_1} \eta_{1i_1}|\xi_{1i_1} \cdots \xi_{ri_r}|$$

$$= \left|\left(\sum_{i_1} \eta_{1i_1}\xi_{1i_1}\right)\xi_{2i_2} \cdots \xi_{ri_r}\right| = |Sx_1y_1, \xi_{2i_2} \cdots \xi_{ri_r}|;$$

again

$$S|x_1x_2 \cdots x_r|\,|y_1y_2e_{i_3} \cdots e_{i_r}| = \sum_{i_2} \eta_{2i_2} S|x_1 \cdots x_r|\,|y_1e_{i_2} \cdots e_{i_r}|$$

$$= \sum_{i_2} \eta_{2i_2}|Sx_1y_1\,\xi_{2i_2} \cdots \xi_{ri_r}|$$

$$= |Sx_1y_1\,Sx_2y_2\,\xi_{3i_3} \cdots \xi_{ri_r}|.$$

The lemma follows easily by a repetition of this process.

The Laplace expansion of a determinant can clearly be expressed as a scalar product. This is most easily done by introducing the notion of the complement of a vector relative to the fundamental basis. If $i_1, i_2, \cdots, i_r$ is a sequence of distinct integers in natural order each less than or equal to n and $i_{r+1}, \cdots, i_n$ the remaining integers up to and including n, also arranged in natural order, the complement of $| e_{i_1}e_{i_2} \cdots e_{i_r} |$ relatively to the fundamental basis is defined as[1]

$$(4) \qquad | e_{i_1}e_{i_2} \cdots e_{i_r} |_c = (-1)^{\Sigma i_\alpha + r(r+1)/2} | e_{i_{r+1}} e_{i_{r+2}} \cdots e_{i_n} |$$

and the complement of $| x_1x_2 \cdots x_r |$ by

$$(5) \qquad | x_1x_2 \cdots x_r |_c = \sum_{(i)}^{*} | \xi_{1i_1}\xi_{2i_2} \cdots \xi_{ri_r} || e_{i_1}e_{i_2} \cdots e_{i_r} |_c,$$

which is a vector of grade $n - r$.

Laplace's expansion of a determinant in terms of minors of order r can now be expressed in the following form.

Lemma 2.

$$(6) \quad S | x_1x_2 \cdots x_r |_c | x_{r+1}x_{r+2} \cdots x_n | = | \xi_{11}\xi_{22} \cdots \xi_{nn} | = | Sx_ie_j |$$
$$= S | x_1 \cdots x_n || e_1 \cdots e_n | = (-1)^{r(n-r)} S | x_1x_2 \cdots x_r || x_{r+1} \cdots x_n |_c.$$

Further as an immediate consequence of (5) we have

Lemma 3.

$$(7) \quad S | x_1x_2 \cdots x_r |_c | y_1y_2 \cdots y_r |_c = S | x_1x_2 \cdots x_r || y_1y_2 \cdots y_r |.$$

5.03 **Compound matrices.** If $A = \Sigma a_{ij}e_{ij}$, then, as in (1),

$$| Ax_1Ax_2 \cdots Ax_r | = \sum_{(j)}^{*} | \xi_{1j_1} \cdots \xi_{rj_r} || Ae_{j_1} \cdots Ae_{j_r} |.$$

But $Ae_j = \sum_i a_{ij}e_i$; so a second application of (1) gives

$$| Ax_1Ax_2 \cdots Ax_r | = \sum_{(i)}^{*} \sum_{(j)}^{*} | \xi_{1j_1} \cdots \xi_{rj_r} || a_{i_1j_1} \cdots a_{i_rj_r} || e_{i_1} \cdots e_{i_r} |.$$

But the determinants $| \xi_{1j_1} \cdots \xi_{rj_r} |$ are the coordinates of the r-vector $| x_1x_2 \cdots x_r |$; hence $| Ax_1 \cdots Ax_r |$ is a linear vector form in $| x_1x_2 \cdots x_r |$ in the corresponding space of $\binom{n}{r}$ dimensions. We denote this vector function or matrix by $C_r(A)$ and write

$$(8) \qquad | Ax_1Ax_2 \cdots Ax_r | = C_r(A) | x_1x_2 \cdots x_r |.$$

We shall call $C_r(A)$ the rth *compound* of A. Important particular cases are

$$(8') \qquad C_1(A) = A, \quad C_n(A) = | A |,$$

[1] The Grassmann notation cannot be conveniently used here since it conflicts with the notation for a determinant. It is sometimes convenient to define the complement of $| e_1e_2 \cdots e_n |$ as 1.

and, if k is a scalar,

(8″) $$C_r(k) = k^r.$$

THEOREM 1.

(9) $$C_r(AB) = C_r(A)C_r(B).$$

For

$$|ABx_1ABx_2 \cdots ABx_r| = C_r(A)\,|Bx_1Bx_2 \cdots Bx_r| = C_r(A)C_r(B)\,|x_1x_2 \cdots x_r|.$$

Corollary. If $|A| \neq 0$, then

(10) $$[C_r(A)]^{-1} = C_r(A^{-1}).$$

THEOREM 2.

(11) $$[C_r(A)]' = C_r(A').$$

For $S\,|x_1x_2 \cdots x_r|\,C_r(A)\,|y_1y_2 \cdots y_r| = |Sx_iAy_j| = |SA'x_iy_j|$

$= S\,|A'x_1 \cdots A'x_r||y_1 \cdots y_r| = S\,|y_1 \cdots y_r|\,C_r(A')\,|x_1 \cdots x_r|.$

THEOREM 3. *If* $A = \sum_1^m a_iSb_i$, *then*

(12) $$C_r(A) = \sum_{(i)}{}^{*}\,|a_{i_1}a_{i_2} \cdots a_{i_r}|\,S\,|b_{i_1}b_{i_2} \cdots b_{i_r}|.$$

This theorem follows by direct substitution for A in the left-hand side of (8). It gives a second proof for Theorem 2.

If $r = m$, (12) consists of one term only, and this term is 0 unless m is the rank of A, a property which might have been made the basis of the definition of rank. In particular, if $X = \sum_1^r e_iSx_i$, $Y = \sum_1^r y_iSe_i$, then $C_r(X) = |e_1e_2 \cdots e_r|\,S\,|x_1x_2 \cdots x_r|$, $C_r(Y) = |y_1y_2 \cdots y_r|\,S\,|e_1e_2 \cdots e_r|$ so that $C_r(XY) = |e_1e_2 \cdots e_r|\,S\,|x_1x_2 \cdots x_r||y_1y_2 \cdots y_r|\,S\,|e_1e_2 \cdots e_r|$. But $XY = \sum_{i,j} e_iSx_iy_jSe_j$ so that $C_r(XY) = |Sx_iy_i||e_1e_2 \cdots e_r|\,S\,|e_1e_2 \cdots e_r|$. Comparing these two forms of $C_r(XY)$ therefore gives another proof of the first lemma of §5.02.

If we consider the complement of $|Ax_1Ax_2 \cdots Ax_r|$ we arrive at a new matrix $C^r(A)$ of order $\binom{n}{r}$ which is called the *rth supplementary compound* of A. From (7) and (12) we have

(13) $$|Ax_1Ax_2 \cdots Ax_r|_c = \sum_i{}^{*}\,|a_{i_1} \cdots a_{i_r}|_c\,S\,|b_{i_1} \cdots b_{i_r}|_c\,|x_1 \cdots x_r|_c = C^r(A)\,|x_1x_2 \cdots x_r|_c$$

and derive immediately the following which are analogous to Theorems 1 and 2.

THEOREM 4.

(14) $$C^r(AB) = C^r(A)C^r(B).$$

THEOREM 5.

(15) $$[C^r(A)]' = C^r(A').$$

The following theorems give the connection between compounds and supplementary compounds and also compounds of compounds.

THEOREM 6.

(16) $$C^r(A')C_{n-r}(A) = |A| = C^{n-r}(A)C_r(A').$$

This is the Laplace expansion of the determinant $|A|$. Using equation (6) and setting $|e|$ for $|e_1e_2 \cdots e_n|$ we have

$$\begin{aligned} |A|\,S\,|x_1x_2 \cdots x_r|_c|x_{r+1} \cdots x_n| &= |A|\,S\,|x_1 \cdots x_n|\,|e| \\ &= S\,|Ax_1 \cdots Ax_n|\,|e| \\ &= S\,|Ax_1 \cdots Ax_r|_c\,|Ax_{r+1} \cdots Ax_n| \\ &= SC^r(A)\,|x_1 \cdots x_r|_cC_{n-r}(A)\,|x_{r+1} \cdots x_n| \\ &= S\,|x_1 \cdots x_r|_cC^r(A')C_{n-r}(A)\,|x_{r+1} \cdots x_n| \end{aligned}$$

and, since the x's are arbitrary, the first part of the theorem follows. The second part is proved in a similar fashion.

Putting $r = n - 1$ in (16) gives the following corollary.

Corollary. $$\text{adj } A = C^{n-1}(A').$$

THEOREM 7.

(17) $$|C_r(A)| = |A|^{\binom{n-1}{r-1}} = |C^r(A)|.$$

For from (16) with A' in place of A, and from the fact that the order of $C_r(A)$ is $\binom{n}{r}$, we have

$$|A|^{\binom{n}{r}} = |C^r(A)C_{n-r}(A')| = |C^r(A)|\,|C_{n-r}(A')|$$

and, since $|A|$ is irreducible when the coordinates of A are arbitrary variables, it follows that $|C^r(A)|$ is a power of $|A|$. Considerations of degree then show that the theorem is true when the coordinates are variables and, since the identity is integral, it follows that it is also true for any particular values of these variables.

THEOREM 8.

(18) $$|A|^{\binom{n-1}{r}} C_s(C_r(A)) = |A|^s C^{\binom{n}{r}-s}(C^{n-r}(A))$$

(19) $$|A|^{\binom{n-1}{r}} C_s(C^r(A)) = |A|^s C^{\binom{n}{r}-s}(C_{n-r}(A)).$$

Using (15), (16) and (17) we get

$$C_s(C^{n-r}(A'))C^{\binom{n}{r}-s}(C^{n-r}(A)) = |C^{n-r}(A)| = |A|^{\binom{n-1}{r}}$$

therefore

$$\begin{aligned} |A|^{\binom{n-1}{r}} C_s(C_r(A)) &= C_s(C_r(A))C_s(C^{n-r}(A'))C^{n-s}(C^{n-r}(A)) \\ &= C_s(C_r(A)C^{n-r}(A'))C^{n-s}(C^{n-r}(A)) \\ &= C_s(|A|)C^{n-s}(C^{n-r}(A)) \\ &= |A|^s C^{n-s}(C^{n-r}(A)). \end{aligned}$$

Similarly

$$C_s(C_{n-r}(A'))C_s(C^r(A)) = C_s(|A|) = |A|^s$$

and therefore

$$\begin{aligned} |A|^s C^{\binom{n}{r}-s}(C_{n-r}(A)) &= C^{\binom{n}{r}-s}(C_{n-r}(A))C_s(C_{n-r}(A'))C_s(C^r(A)) \\ &= |C_{n-r}(A)| C_s(C^r(A)) \\ &= |A|^{\binom{n-1}{r}} C_s(C^r(A)). \end{aligned}$$

An important particular case is $C_s(C^{n-1}(A)) = |A|^{s-1}C^{n-s}(A)$ whence

(20) $$C_s(\text{adj } A) = C_s(C^{n-1}(A')) = |A|^{s-1}C^{n-s}(A').$$

5.04 **Roots of compound matrices.** If A has simple elementary divisors and its roots are $\lambda_1, \lambda_2, \cdots, \lambda_n$, the corresponding invariant vectors being $a_1, a_2, \cdots, a_n$, then the roots of $C_r(A)$ are the products $\lambda_{i_1}\lambda_{i_2} \cdots \lambda_{i_r}$ in which no two subscripts are the same and the subscripts are arranged in, say, numerical order; and the invariant vector corresponding to $\lambda_{i_1}\lambda_{i_2} \cdots \lambda_{i_r}$ is $|a_{i_1}a_{i_2} \cdots a_{i_r}|$. For there are $\binom{n}{r}$ distinct vectors of this type and

$$C_r(A)|a_{i_1}a_{i_2} \cdots a_{i_r}| = |Aa_{i_1} \cdots Aa_{i_r}| = \lambda_{i_1}\lambda_{i_2} \cdots \lambda_{i_r}|a_{i_1}a_{i_2} \cdots a_{i_r}|.$$

Similarly for $C^r(A)$ the roots and invariants are $\lambda_{i_1}\lambda_{i_2} \cdots \lambda_{i_r}$ and $|a_{i_1}a_{i_2} \cdots a_{i_r}|_c$.

It follows from considerations of continuity that the roots are as given above even when the elementary divisors are not simple.

5.05 **Bordered determinants.** Let $A = \|a_{ij}\| = \sum_{j=1}^{n} a_j S e_j$, $a_j = \sum_i a_{ij}e_i$, be any matrix and associate with it two sets of vectors

$$X\colon x_i = \sum_{j=1}^{n} \xi_{ij}e_j, \qquad (i = 1, 2, \cdots, r)$$

$$Y\colon y_i = \sum_{j=1}^{n} \eta_{ij}e_j.$$

Consider the bordered determinant

$$
(21) \qquad \Delta_r = \begin{vmatrix} a_{11} & a_{12} & \cdots & a_{1n} & \xi_{11} & \cdots & \xi_{r1} \\ a_{21} & a_{22} & \cdots & a_{2n} & \xi_{12} & \cdots & \xi_{r2} \\ \cdots & \cdots & \cdots & \cdots & \cdots & \cdots & \cdots \\ \cdots & \cdots & \cdots & \cdots & \cdots & \cdots & \cdots \\ a_{n1} & a_{n2} & \cdots & a_{nn} & \xi_{1n} & \cdots & \xi_{rn} \\ \eta_{11} & \eta_{12} & \cdots & \eta_{1n} & 0 & \cdots & 0 \\ \cdots & \cdots & \cdots & \cdots & \cdots & \cdots & \cdots \\ \eta_{r1} & \eta_{r2} & \cdots & \eta_{rn} & 0 & \cdots & 0 \end{vmatrix} = \begin{vmatrix} A & X \\ Y' & 0_r \end{vmatrix}
$$

where $r < n$, and 0_r is a square block of 0's with r rows and columns.

If we introduce r additional fundamental units $e_{n+1}, \cdots, e_{n+r}$, Δ_r can be regarded as the determinant of a matrix $\mathfrak{A}$ of order $n + r$, namely,

$$\mathfrak{A} = \sum_1^n a_i S e_i + \sum_1^r x_i S e_{n+i} + \sum_1^r e_{n+i} S y_i = \sum_1^{n+2r} c_i S d_i.$$

If now we form $| \mathfrak{A} | = S | e | C_{n+r}(\mathfrak{A}) | e |$ as in §5.03, we have

$$C_{n+r}(\mathfrak{A}) = \sum_{(i)}^{*} | c_{i_1} \cdots c_{i_{n+r}} | S | d_{i_1} \cdots d_{i_{n+r}} | \quad (i = 1, 2, \cdots, n + 2r).$$

In this form any $| c_{i_1} \cdots c_{i_{n+r}} |$ which contains more than n out of $a_1, \cdots, a_n$, $x_1, \cdots, x_r$ is necessarily 0; also, if it does not contain all the x's, the corresponding $| d_{i_1}, \cdots, d_{i_{n+r}} |$ will contain more than n out of $e_1, \cdots, e_n$, $y_1, \cdots, y_r$ and is consequently 0. We therefore have

$$C_{n+r}(\mathfrak{A}) = \sum_{(i)}^{*} | a_{i_1} a_{i_2} \cdots a_{i_{n-r}} x_1 x_2 \cdots x_r e_{n+1} \cdots e_{n+r} |$$
$$\times S | e_{i_1} e_{i_2} \cdots e_{i_{n-r}} y_1 y_2 \cdots y_r e_{n+1} \cdots e_{n+r} | \quad (i = 1, 2, \cdots, n)$$

and hence, passing back to space of n dimensions,

$$| \mathfrak{A} | = \sum_i^{*} S | e || a_{i_1} \cdots a_{i_{n-r}} x_1 \cdots x_r | S | e_{i_1} \cdots e_{i_{n-r}} y_1 \cdots y_r || e |$$
$$= \Sigma^* S | x_1 \cdots x_r || a_{i_1} \cdots a_{i_{n-r}} |_c S | e_{i_1} \cdots e_{i_{n-r}} |_c | y_1 \cdots y_r |$$
$$= S | x_1 \cdots x_r | C^{n-r}(A) | y_1 \cdots y_r |.$$

This relation shows why the bordered determinant is frequently used in place of the corresponding compound in dealing with the theory of forms.

5.06 The reduction of bilinear forms. The Lagrange method of reducing quadratic and bilinear forms to a normal form is, as we shall now see, closely connected with compounds.

If A is any matrix, not identically 0, there exist vectors x_1, y_1 such that $Sx_1Ay_1 \neq 0$; then, setting $A = A_1$ for convenience, the matrix

$$A_2 = A_1 - A_1y_1 \frac{SA_1'x_1}{Sx_1A_1y_1}$$

has its rank exactly 1 less than that of A. For, if $A_1z = 0$, then

$$A_2z = A_1z - A_1y_1 \frac{SA'x_1 \cdot z}{Sx_1A_1y_1} = A_1z - A_1y_1 \frac{Sx_1A_1z}{Sx_1A_1y_1} = 0$$

and, conversely if $A_2z = 0$, then

$$A_1z = A_1y_1 \frac{Sx_1A_1z}{Sx_1A_1y_1} = kA_1y_1,$$

say, or $A_1(z - ky_1) = 0$. The null-space of A_2 is therefore obtained from that of A_1 by adding y_1 to its basis, which increases the order of this space by 1 since $A_1y_1 \neq 0$.

If $A_2 \neq 0$, this process may be repeated, that is, there exist x_2, y_2 such that $Sx_2A_2y_2 \neq 0$ and the rank of

$$A_3 = A_2 - A_2y_2 \frac{SA_2'x_2}{Sx_2A_2y_2}$$

is 1 less than that of A_2. If the rank of A is r, we may continue this process by setting

(22) $$A_{s+1} = A_s - A_sy_s \frac{SA_s'x_s}{Sx_sA_sy_s} \qquad (s = 1, 2, \cdots, r)$$

where $Sx_sA_sy_s \neq 0$ and $A_1 = A$, $A_{r+1} = 0$; we then have

(23) $$A = \sum_{s=1}^{r} A_sy_s \frac{SA_s'x_s}{Sx_sA_sy_s} = \sum_{1}^{r} \mathfrak{A}_s$$

where $\mathfrak{A}_s = A_sy_s \dfrac{SA_s'x_s}{Sx_sA_sy_s}$ is a matrix of rank 1. Generally speaking, one may take $x_s = y_s$ and it is of some interest to determine when this is not possible. If $SxBx = 0$ for every x, we readily see that B is skew. For then $Se_iBe_i = Se_jBe_j = S(e_i + e_j)B(e_i + e_j) = 0$ and therefore

$$0 = S(e_i + e_j)B(e_i + e_j) = Se_iBe_i + Se_jBe_j + Se_iBe_j + Se_jBe_i,$$

that is, $Se_iBe_j = -Se_jBe_i$ and hence $B' = -B$. Hence we may take $x_s = y_s$ so long as $A_s \neq -A_s'$.

5.07 We shall now derive more explicit forms for the terms in (23) and show how they lead to the Sylvester-Francke theorems on compound determinants.

Let $x^1, x^2, \cdots, x^r, y^1, y^2, \cdots, y^r$ be variable vectors and set

(24) $$\begin{aligned} J &= S \,|\, x_sx^1x^2 \cdots x^r \,|\, C_{r+1}(A_s) \,|\, y_sy^1y^2 \cdots y^r \,| \\ &= S \,|\, x_sx^1x^2 \cdots x^r \,|\,|\, A_sy_sA_sy^1 \cdots A_sy^r \,|; \end{aligned}$$

then from (22)

$$J = S \mid x_s x^1 \cdots x^r \mid\mid A_s y_s A_{s+1} y^1 \cdots A_{s+1} y^r \mid$$
$$= \mid S x_s A_s y_s S x^1 A_{s+1} y^1 \cdots S x^r A_{s+1} y^r \mid.$$

If the x's denote rows in this determinant, the first row is

$$S x_s A_s y_s,\ S x_s A_{s+1} y^1,\ \cdots,\ S x_s A_{s+1} y^r$$

each term of which is 0 except the first, since x_s lies in the null-space of A'_{s+1}, and $S x_s A_s y_s \neq 0$. Hence

(25) $$J = S x_s A_s y_s \mid S x^1 A_{s+1} y^1 \cdots S x^r A_{s+1} y^r \mid$$

and therefore from (24)

(26) $$S \mid x_s x^1 \cdots x^r \mid C_{r+1}(A_s) \mid y_s y^1 \cdots y^r \mid$$
$$= S x_s A_s y_s S \mid x^1 \cdots x^r \mid C_r(A_{s+1}) \mid y^1 \cdots y^r \mid.$$

Repeated application of this relation gives

(27) $$S \mid x_s x_{s+1} \cdots x_{s+t-1} x^1 x^2 \cdots x^r \mid C_{r+t}(A_s) \mid y_s \cdots y_{s+t-1} y^1 \cdots y^r \mid$$
$$= S x_s A_s y_s S x_{s+1} A_{s+1} y_{s+1} \cdots S x_{s+t-1} A_{s+t-1} y_{s+t-1} S \mid x^1 \cdots x^r \mid$$
$$\cdot C_r(A_{s+t}) \mid y^1 \cdots y^r \mid,$$

a particular case of which is

(27′) $$S \mid x_1 x_2 \cdots x_{s-1} x \mid C_s(A) \mid y_1 \cdots y_{s-1} y \mid$$
$$= S x_1 A_1 y_1 \cdots S x_{s-1} A_{s-1} y_{s-1} S x A_s y.$$

To simplify these and similar formulae we shall now use a single letter to indicate a sequence of vectors; thus we shall set $X_{s,\,s+t-1}$ for $x_s x_{s+1} \cdots x_{s+t-1}$ and Y^r for $y^1 y^2 \cdots y^r$; also $C_{r,\,s}$ for $C_r(A_s)$. Equations (26) and (27) may then be written

(26a) $$S \mid x_s X^r \mid C_{r+1,\,s} \mid y_s Y^r \mid = S x_s A_s y_s S \mid X^r \mid C_{r,\,s+1} \mid Y^r \mid,$$

(27a) $$S \mid X_{s,\,s+t-1} X^r \mid C_{r+t,\,s} \mid Y_{s,\,s+t-1} Y^r \mid = \prod_{i=s}^{s+t-1} S x_i A_i y_i S \mid X^r \mid C_{r,\,s+t} \mid Y^r \mid.$$

We get a more convenient form for (26a), namely

(28) $$S \mid X_{s,\,t} X^r \mid C_{r+t-s+1,\,s} \mid Y_{s,\,t} Y^r \mid$$
$$= S x_s A_s y_s S \mid X_{s+1,\,t} X^r \mid C_{r+t-s,\,s+1} \mid Y_{s+1,\,t} Y^r \mid$$

by replacing r by $r + t - s$ and then changing $x^1 x^2 \cdots x^{r+t-s}$ into $x_{s+1} \cdots x_t x^1 \cdots x^r$ along with a similar change in the y's. Putting $s = 1, 2, \cdots, t$ in succession and forming the product of corresponding sides of the equations so obtained from (28) we get after canceling the common factors, which are not identically 0 provided that $r + t$ is not greater than the rank of A,

$$(29)\qquad S \mid X_tX^r \mid C_{r+t,1} \mid Y_tY^r \mid = \prod_1^t Sx_iA_iy_i \cdot S \mid X^r \mid C_{r,t+1} \mid Y^r \mid,$$

or from (27′)

$$(30)\qquad S \mid X_tX^r \mid C_{r+t} \mid Y_tY^r \mid = S \mid X_t \mid C_t \mid Y_t \mid S \mid X^r \mid C_{r,t+1} \mid Y^r \mid$$

which may also be written in the form

$$(30')\qquad K \equiv \frac{S \mid X_tX^r \mid C_{r+t} \mid Y_tY^r \mid}{S \mid X_t \mid C_t \mid Y_t \mid} = \mid Sx^iA_{t+1}y^j \mid;$$

in particular

$$(31)\qquad \frac{S \mid X_tx \mid C_{t+1}(A) \mid Y_ty \mid}{S \mid X_t \mid C_t(A) \mid Y_t \mid} = SxA_{t+1}y.$$

This gives a definition of A_{t+1} which may be used in place of (22); it shows that this matrix depends on $2t$ vector parameters. It is more convenient for some purposes to use the matrix $A^{(t)}$ defined by

$$(32)\qquad SxA^{(t)}y = S \mid X_tx \mid C_{t+1}(A) \mid Y_ty \mid.$$

From (31) we then have $Sx^iA_{t+1}y^j = Sx^iA^{(t)}y^j/S \mid X_t \mid C_t \mid Y_t \mid$ and therefore from (30′)

$$(33)\qquad K = \frac{\mid Sx^iA^{(t)}y^j \mid}{[S \mid X_t \mid C_t \mid Y_t \mid]^r} = \frac{S \mid X^r \mid C_r(A^{(t)}) \mid Y^r \mid}{[S \mid X_t \mid C_t(A) \mid Y_t \mid]^r}.$$

Hence

$$(34)\qquad S \mid X_tX^r \mid C_{r+t}(A) \mid Y_tY^r \mid = \frac{S \mid X^r \mid C_r(A^{(t)}) \mid Y^r \mid}{[S \mid X_t \mid C_t(A) \mid Y_t \mid]^{r-1}}$$

which is readily recognized as Sylvester's theorem if the x's are replaced by fundamental units and the integral form of (33) is used.

5.08 Invariant factors.

We shall now apply the above results in deriving the normal form of §3.02. We require first, however, the following lemma.

LEMMA 4. *If $A(\lambda)$ is a matric polynomial, there exists a constant vector y and a vector polynomial x such that $SxAy$ is the highest common factor of the coordinates of A.*

Let $y = \Sigma\eta_ie_i$ be a vector whose coordinates are variables independent of λ. Let α_1 be the H. C. F. of the coordinates of $A = \| a_{ij} \|$ and set

$$A = \alpha_1B, \quad By = \Sigma\eta_jb_{ij}e_j = \Sigma\beta_ie_i.$$

There is no value λ_1 of λ independent of the η's for which every $\beta_i = 0$; for if this were so, $\lambda - \lambda_1$ would be a factor of each b_{ij} and α_1 could not then be the H. C. F. of the a_{ij}. Hence the resultant of $\beta_1, \beta_2, \cdots, \beta_n$ as polynomials in λ is not identically 0 as a polynomial in the η's; there are therefore values of the η's for which this resultant is not 0, and for these values the β's have no factor common to all. There then exist scalar polynomials $\xi_1, \xi_2, \cdots, \xi_n$ such that $\Sigma\xi_i\beta_i = 1$ and therefore, if $x = \Sigma\xi_i e_i$, we have $SxBy = 1$ or $SxAy = \alpha_1$.

Returning now to the form of A given in §5.06, namely

$$A = \sum_1^r \frac{A_s y_s SA'_s x_s}{Sx_s A_s y_s},$$

we can as above choose x_s, y_s in such a manner that $Sx_sA_sy_s = \alpha_s$ is the highest common factor of the coordinates of A_s and, when this is done, $v_s = A_s y_s/\alpha_s$, $u_s = A'_s x_s/\alpha_s$ are integral in λ. We then have

$$(35) \qquad A = \sum_1^r \frac{A_s y_s SA'_s x_s}{\alpha_s} = \Sigma\alpha_s v_s Su_s.$$

Moreover $A_s y_i = 0 = A'_s x_i$ when $i < s$ and therefore in

$$S\,|\,x_1 \cdots x_r\,|\,|\,A_1y_1A_2y_2 \cdots A_ry_r\,| = |\,Sx_iA_jy_j\,| = |\,SA'_jx_iy_j\,|$$

all terms on one side of the main diagonal are 0 so that it reduces to $Sx_1A_1y_1 \cdots Sx_rA_ry_r = \alpha_1\alpha_2 \cdots \alpha_r$. Hence, dividing by $\alpha_1 \cdots \alpha_r$ and replacing A_jy_j/α_j by v_j as above, we see that $|\,x_1 \cdots x_r\,|$ and $|\,v_1 \cdots v_r\,|$ are not 0 for any value of λ, and therefore the constituent vectors in each set remain linearly independent for all values of λ. It follows in the same way that the sets $u_1, \cdots, u_r$ and $y_1, \cdots, y_r$, respectively, are also linearly independent for all values of λ, that is, these four sets are elementary sets. It follows from Theorem 5 §4.03, that we can find elementary polynomials P and Q such that

$$Pv_i = e_i = Q'u_i \quad (i = 1, 2, \cdots, r),$$

and hence

$$(36) \qquad PAQ = P\left(\sum_1^r \alpha_s v_s Su_s\right)Q = \sum_1^r \alpha_s e_s Se_s,$$

which is the normal form of §3.02.

5.09 Vector products. Let $x_i = \Sigma\xi_{ij}e_j$ $(i = 1, 2, \cdots, r)$ be a set of arbitrary vectors and consider the set of all products of the form $\xi_{1i_1}\xi_{2i_2} \cdots \xi_{ri_r}$ arranged in some definite order. These products may then be regarded as the coordinates of a hypernumber[2] of order n^r which we shall call the *tensor product*[3]

[2] The term 'hypernumber' is used in place of vector, as defined in §1.01 since we now wish to use the term 'vector' in a more restricted sense.

[3] This product was called by Grassmann the general or indeterminate product.

of $x_1, x_2, \cdots, x_r$ and we shall denote it by $x_1x_2 \cdots x_r$. In particular if we take all the products $e_{i_1}e_{i_2} \cdots e_{i_r}$ $(i_1, i_2, \cdots, i_r = 1, 2, \cdots, n)$ each has all its coordinates zero except one, which has the value 1, and no two are equal. Further

$$x_1x_2 \cdots x_r = \Sigma\xi_{1i_1}\xi_{2i_2} \cdots \xi_{ri_r}e_{i_1}e_{i_2} \cdots e_{i_r}.$$

If we regard the products $e_{i_1}e_{i_2} \cdots e_{i_r}$ as the basis of the set of hypernumbers, we are naturally led to consider sums of the type

$$w = \Sigma\omega_{i_1i_2 \cdots i_r}e_{i_1}e_{i_2} \cdots e_{i_r}$$

where the ω's are scalars; and we shall call such a hypernumber a *tensor* of *grade* r. It is readily seen that the product $x_1x_2 \cdots x_r$ is distributive and homogeneous with regard to each of its factors, that is,

$$x_1(\lambda x_2 + \mu y_2)x_3 \cdots x_r = \lambda x_1x_2 \cdots x_r + \mu x_1y_2x_3 \cdots x_r.$$

The product of two tensors of grade r and s is then defined by assuming the distributive law and setting

$$(e_{i_1}e_{i_2} \cdots e_{i_r})(e_{j_1}e_{j_2} \cdots e_{j_s}) = e_{i_1} \cdots e_{i_r}e_{j_1} \cdots e_{j_s}.$$

It is easily shown that the product so defined is associative; it is however not commutative as is seen from the example

$$x_1x_2 - x_2x_1 = \Sigma\Sigma(\xi_{1i_1}\xi_{2i_2} - \xi_{1i_2}\xi_{2i_1})e_{i_1}e_{i_2}$$

$$= \sum_{(i)}{}^{*} \begin{vmatrix} \xi_{1i_1} & \xi_{1i_2} \\ \xi_{2i_1} & \xi_{2i_2} \end{vmatrix} (e_{i_1}e_{i_2} - e_{i_2}e_{i_1}).$$

Here the coefficients of $e_{i_1}e_{i_2} - e_{i_2}e_{i_1}$ $(i_1 < i_2)$ are the coordinates of $|x_1x_2|$ so that this tensor might have been defined in terms of the tensor product by setting

$$|x_1x_2| = x_1x_2 - x_2x_1.$$

In the same way, if we form the expression[4]

$$f(x_1, x_2, \cdots, x_r) = \Sigma \operatorname{sgn}(i_1, i_2, \cdots, i_r)x_{i_1}x_{i_2} \cdots x_{i_r},$$

and expand it in terms of the coordinates of the x's and the fundamental units, it is readily shown that the result is

$$\sum_{(i)}{}^{*} |\xi_{1i_1}\xi_{2i_2} \cdots \xi_{ri_r}| f(e_{i_1}, e_{i_2}, \cdots, e_{i_r}).$$

[4] The determinant of a square array of vectors x_{ij} $(i, j = 1, 2, \cdots, r)$ may be defined as

$$|x_{ij}| = \Sigma \operatorname{sgn}(i_1, i_2, \cdots, i_r)x_{1i_1}x_{2i_2} \cdots x_{ri_r}.$$

In this definition the row marks are kept in normal order and the column marks permuted; a different expression is obtained if the rôles of the row and column marks are interchanged but, as these determinants seem to have little intrinsic interest, it is not worth while to develop a notation for the numerous variants of the definition given above.

Here the scalar multipliers are the same as the coordinates of $| x_1x_2 \cdots x_r |$ and hence the definition of §1.11 may now be replaced by

$$| x_1x_2 \cdots x_r | = \Sigma \text{sgn}(i_1, i_2, \cdots, i_r)x_{i_1}x_{i_2} \cdots x_{i_r},$$

which justifies the notation used. We then have

$$(37) \qquad | x_1x_2 \cdots x_r | = \sum_{(i)}^{*} | \xi_{1i_1}\xi_{2i_2} \cdots \xi_{ri_r} | \, | e_{i_1}e_{i_2} \cdots e_{i_r} |.$$

It is easily seen that the tensors $| e_{i_1}e_{i_2} \cdots e_{i_r} |$ are linearly independent and (37) therefore shows that they form a basis for the set of vectors of grade r. Any expression of the form

$$\Sigma \xi_{i_1i_2 \ldots i_r} | e_{i_1}e_{i_2} \cdots e_{i_r} |$$

is called a vector of grade r and a vector of the form (37) is called a pure vector of grade r.

5.10 **The direct product.** If $A_i = \| a_{pq}^{(i)} \|$ $(i = 1, 2, \cdots, r)$ is a sequence of matrices of order n, then

$$(38) \qquad A_1x_1A_2x_2 \cdots A_rx_r = \sum_{i,j} a_{i_1j_1}^{(1)} a_{i_2j_2}^{(2)} \cdots a_{i_rj_r}^{(r)} \xi_{1j_1}\xi_{2j_2} \cdots \xi_{rj_r} e_{i_1}e_{i_2} \cdots e_{i_r}$$
$$= \mathfrak{A}(x_1x_2 \cdots x_r)$$

where $\mathfrak{A}$ is a linear homogeneous tensor function of $x_1x_2 \cdots x_r$, that is, a matrix in space of n^r dimensions. This matrix is called the direct product[5] of $A_1, A_2, \cdots, A_r$ and is denoted by $A_1 \times A_2 \times \cdots \times A_r$. Obviously

$$(39) \qquad A_1B_1 \times A_2B_2 \times \cdots = (A_1 \times A_2 \times \cdots)(B_1 \times B_2 \times \cdots),$$

and the form of (38) shows that

$$(40) \qquad (A_1 \times A_2 \times \cdots)' = A_1' \times A_2' \times \cdots.$$

From (39) we have, on putting $r = 1$ for convenience,

$$A_1 \times A_2 \times A_3 = (A_1 \times 1 \times 1)(1 \times A_2 \times 1)(1 \times 1 \times A_3).$$

Making $A_i = 1$ $(i = 2, 3, \cdots, r)$ in (38) we have

$$A_1x_1x_2 \cdots, x_r = \Sigma a_{i_1j_1}^{(1)} \xi_{1j_1}\xi_{2i_2} \cdots \xi_{ri_r} e_{i_1}e_{i_2} \cdots e_{i_r}$$

and hence the determinant of the corresponding matrix equals $| A_1 |^{n^{r-1}}$. Treating the other factors in the same way we then see that

$$(41) \qquad | A_1 \times A_2 \times \cdots \times A_r | = | A_1A_2 \cdots A_r |^{n^{r-1}}.$$

Again if as in §5.04 we take x_1 as an invariant vector of A_1, x_2 as an invariant vector of A_2, and so on, and denote the roots of A_i by λ_{ij}, we see that the roots

[5] This definition may be generalized by taking $x_1, x_2, \cdots$ as vectors in different spaces of possibly different orders. See also §7.03.

of $A_1 \times A_2 \times \cdots \times A_r$ are the various products $\lambda_{1j_1}\lambda_{2j_2} \cdots \lambda_{rj_r}$. When the roots of each matrix are distinct, this gives equation (41) and, since this is an integral relation among the coefficients of the A's, it follows that it is true in general.

An important particular case arises when each of the matrices in (38) equals the same matrix A; the resultant matrix is denoted by $\Pi_r(A)$, that is

$$\Pi_r(A) = A \times A \times \cdots \quad (r \text{ factors}). \tag{42}$$

It is sometimes called the *product transformation*. Relations (39), (40), and (41) then become

$$\Pi_r(AB) = \Pi_r(A)\Pi_r(B),\ \Pi_r(A)' = \Pi_r(A'),\ |\,\Pi_r(A)\,| = |\,A\,|^{rn^{r-1}}. \tag{43}$$

5.11 Induced or power matrices. If $x_1, x_2, \cdots, x_r$ are arbitrary vectors, the symmetric expression obtained by forming their products in every possible order and adding is called a *permanent*. It is usually denoted by $\overset{+}{|}\, x_1x_2 \cdots x_r \,\overset{+}{|}$ but it will be more convenient here to denote it by $\{x_1x_2 \cdots x_r\}$; and similarly, if α_{ij} is a square array of scalars, we shall denote by $\{\alpha_{11}\alpha_{22} \cdots \alpha_{rr}\}$ the function $\Sigma\alpha_{1i_1}\alpha_{2i_2} \cdots \alpha_{ri_r}$ in which the summation stretches over every permutation of $1, 2, \cdots, r$.

If some of the x's are equal, the terms of $\{x_1x_2 \cdots x_r\}$ become equal in sets each of which has the same number of terms. If the x's fall into s groups of $i_1, i_2, \cdots, i_s$ members, respectively, the members in each group being equal to one another, then

$$\frac{\{x_1x_2 \cdots x_r\}}{i_1!\, i_2! \cdots i_s!} \qquad (\Sigma i_j = r)$$

has integral coefficients. For the present we shall denote this expression by $\{x_1x_2 \cdots x_r\}^*$, but sometimes it will be more convenient to use the more explicit notation

$$\begin{Bmatrix} x_1 & x_2 & \cdots & x_s \\ i_1 & i_2 & \cdots & i_s \end{Bmatrix}$$

in which i_1 of the x's equal x_1, i_2 equal x_2, etc.; this notation is, in fact, that already used in §2.08, for instance,

$$\{x \quad x \quad y\} = 2x^2y + 2xyx + 2yx^2$$

$$\begin{Bmatrix} x & y \\ 2 & 1 \end{Bmatrix} = x^2y + xyx + yx^2 = \tfrac{1}{2}\{xxy\}.$$

The same convention applies immediately to $\{\alpha_{11}\alpha_{22} \cdots \alpha_{rr}\}$.

In the notation just explained we have

$$\{x_1x_2 \cdots x_r\} = \Sigma''\{\xi_{1i_1}\xi_{2i_2} \cdots \xi_{ri_r}\}^*\{e_{i_1}e_{i_2} \cdots e_{i_r}\} \tag{44}$$

where the summation Σ'' extends over all combinations $i_1i_2 \cdots i_r$ of the first n integers repetition being allowed. This shows that the set of all permanents of grade r has the basis $\{e_{i_1}e_{i_2} \cdots e_{i_r}\}$ of order $(n + r - 1)!/r!(n - 1)!$. From (44) we readily derive

$$(45)\quad \{Ax_1Ax_2 \cdots Ax_r\} = \sum_{i,j}{}'' \{a_{i_1j_1}a_{i_2j_2} \cdots a_{i_rj_r}\}^*\{\xi_{1j_1} \cdots \xi_{rj_r}\}^*\{e_{i_1} \cdots e_{i_r}\}$$

which is a linear tensor form in $\{x_1x_2 \cdots x_r\}$. We may therefore set

$$(46)\qquad \{Ax_1Ax_2 \cdots Ax_r\} = P_r(A)\{x_1x_2 \cdots x_r\},$$

where $P_r(A)$ is a matrix of order $(n + r - 1)!/r!(n - 1)!$ whose coordinates are the polynomials in the coordinates of A which are given in (45); this matrix is called the rth *induced* or *power* matrix of A. As with $C_r(A)$ and $\Pi_r(A)$ it follows that

$$(47)\qquad \begin{gathered} P_r(AB) = P_r(A)P_r(B), \quad P_r(A)' = P_r(A'), \\ |P_r(A)| = |A|^{\binom{n+r-1}{r-1}}; \end{gathered}$$

also the roots of $P_r(A)$ are the various products of the form $\lambda_1^{\alpha_1}\lambda_2^{\alpha_2} \cdots \lambda_r^{\alpha_r}$ for which $\Sigma\alpha_i = r$.

5.12 **Associated matrices.** The matrices considered in the preceding sections have certain common properties; the coordinates of each are functions of the variable matrix A and, if $T(A)$ stands for any one of them, then

$$(48)\qquad T(AB) = T(A)T(B).$$

Following Schur, who first treated the general problem of determining all such matrices, we shall call any matrix with these properties an *associated* matrix. If S is any constant matrix in the same space as $T(A)$, then $T_1(A) = ST(A)S^{-1}$ is clearly also an associated matrix; associated matrices related in this manner are said to be *equivalent*.

Let the orders of A and $T(A)$ be n and m respectively and denote the corresponding identity matrices by 1_n and 1_m; then from (48)

$$(49)\qquad T^2(1_n) = T(1_n), \qquad T(1_n)T(A) = T(A) = T(A)T(1_n).$$

If s is the rank of $T(1_n)$, we can find a matrix S which transforms $T(1_n)$ into a diagonal matrix with s 1's in the main diagonal and zeros elsewhere; and we may without real loss of generality assume that $T(1_n)$ has this form to start with, and write

$$T(1_n) = \begin{Vmatrix} 1_s & 0 \\ 0 & 0 \end{Vmatrix}.$$

The second equation of (49) then shows that $T(A)$ has the form

$$T(A) = \begin{Vmatrix} T_s(A) & 0 \\ 0 & 0 \end{Vmatrix}$$

and we shall therefore assume that $s = m$ so that $T(1_n) = 1_m$. It follows from this that $| T(A) | \neq 0$ so that $T(A)$ is not singular for every A; we shall then say that T is non-singular.

A non-singular associated matrix $T(A)$ is reducible (cf. §3.10) if it can be expressed in the form $T(A) = T_1(A) + T_2(A)$ where, if $E_1 = T_1(1_n)$, $E_2 = T_2(1_n)$, so that $E_1 + E_2 = 1_m$, then

$$T_1(A) = E_1T(A)E_1, \quad T_2(A) = E_2T(A)E_2$$
$$E_1T(A)E_2 = 0 = E_2T(A)E_1$$

so that

$$E_1^2 = E_1, E_2^2 = E_2$$
$$E_1E_2 = 0 = E_2E_1$$

and there is therefore an equivalent associated matrix $t(A)$ which has the form

$$t(A) = \left\| \begin{matrix} t_1(A) & 0 \\ 0 & t_2(A) \end{matrix} \right\|.$$

When $T(A)$ is reducible in this manner we have

$$\begin{aligned} T_1(AB) &= E_1T(AB)E_1 = E_1T(A)T(B)E_1 \\ &= E_1T(A)(E_1 + E_2)T(B)E_1 \\ &= E_1T(A)E_1T(B)E_1 = T_1(A)T_1(B) \end{aligned}$$

so that $T_1(A)$ and $T_2(A)$ are separately associated matrices. We may therefore assume $T(A)$ irreducible without loss of generality since reducible associated matrices may be built up out of irreducible ones by reversing the process used above.

5.13 We shall now show that, if λ is a scalar variable, then $T(\lambda)$ is a power of λ. To begin with we shall assume that the coordinates of $T(\lambda)$ are rational functions in λ and that $T(1)$ is finite; we can then set $T(\lambda) = T_1(\lambda)/f(\lambda)$ where $f(\lambda)$ is a scalar polynomial whose leading coefficient is 1 and the coordinates of $T_1(\lambda)$ are polynomials whose highest common factor has no factor in common with $f(\lambda)$. If μ is a second scalar variable, (48) then gives

$$\frac{T_1(\lambda)T_1(\mu)}{f(\lambda)f(\mu)} = \frac{T_1(\lambda\mu)}{f(\lambda\mu)},$$

hence $f(\lambda\mu)$ is a factor of $f(\lambda)f(\mu)$, from which it follows readily that $f(\lambda\mu) = f(\lambda)f(\mu)$; so that $f(\lambda)$ is a power of λ and also

$$T_1(\lambda\mu) = T_1(\lambda)T_1(\mu). \tag{50}$$

We also have $f(1) = 1$ and hence $T_1(1_n) = T(1_n) = 1_m$.

Let $T_1(\lambda) = F_0 + \lambda F_1 + \cdots + \lambda^s F_s (F_s \neq 0)$; then from (50)

$$F_0 + \lambda\mu F_1 + \cdots + \lambda^s\mu^s F_s = (F_0 + \lambda F_1 + \cdots)(F_0 + \mu F_1 + \cdots)$$

which gives

$$F_i^2 = F_i, F_iF_j = 0 \ (i \neq j), \qquad (i, j = 0, 1, \cdots, s).$$

Now

$$T_1(\lambda)T(A) = f(\lambda)T(\lambda)T(A) = f(\lambda)T(\lambda A) = T(A)T_1(\lambda);$$

therefore

$$\Sigma F_i T(A)\lambda^i = \Sigma T(A)F_i\lambda^i$$

and hence on comparing powers of λ we have

$$F_iT(A) = T(A)F_i$$

and, since $\Sigma F_i = T_1(1) = 1_m$ and we have assumed that $T(A)$ is irreducible, it follows that every $F_i = 0$ except F_s, which therefore equals 1_m. Hence $T_1(\lambda) = \lambda^s$ and, since $f(\lambda)$ is a power of λ, we may set

(51) $$T(\lambda) = \lambda^r.$$

Since $T(\lambda A) = T(\lambda)T(A) = \lambda^r T(A)$, we have the following theorem.

THEOREM 9. *If $T(A)$ is irreducible, and if $T(\lambda)$ is a rational function of the scalar variable λ, then $T(\lambda) = \lambda^r$ and the coordinates of $T(A)$ are homogeneous functions of order r in the coordinates of A.*

The restriction that $T(\lambda)$ is rational in λ is not wholly necessary. For instance, if q is any whole number and ϵ a corresponding primitive root of 1, then $T^q(\epsilon) = 1_m$ and from this it follows without much difficulty that $T(\epsilon) = \epsilon^s$ where s is an integer which may be taken to be the same for any *finite* number of values of q. It follows then that, if $T(\lambda) = \| t_{ij}(\lambda) \|$, the functions $t_{ij}(\lambda)$ satisfy the equation

$$t_{ij}(\epsilon\lambda) = \epsilon^s t_{ij}(\lambda)$$

and under very wide assumptions as to the nature of the functions t_{ij} it follows from this that $T(\lambda)$ has the form λ^r. Again, if we assume that $T(\lambda) = \lambda^\alpha \sum_{-\alpha}^{\alpha} T_r\lambda^r$, then $T(\lambda)T(\mu) = T(\lambda\mu)$ gives immediately

$$T_r\mu^{r+\alpha} = T(\mu)$$

so that only one value of r is admissible and for this value $T_r = 1$ as before.

5.14 If the coordinates of $T(A)$ are rational functions of the coordinates a_{ij} of A, so that r is an integer, we can set $T(A) = T_1(A)/f(A)$ where the coordinates of $T_1(A)$ are integral in the a_{ij} and $f(A)$ is a scalar polynomial in these variables which has no factor common to all the coordinates of $T_1(A)$. As in (50) we then have

$$T_1(AB) = T_1(A)T_1(B), f(AB) = f(A)f(B).$$

It follows from the theory of scalar invariants that $f(A)$ can be taken as a positive integral power of $|A|$; we shall therefore from this point on assume that the coordinates of $T(A)$ are homogeneous polynomials in the coordinates of A unless the contrary is stated explicitly. We shall call r the *index* of $T(A)$.

THEOREM 10. *If $T(A)$ is an associated matrix of order m and index r, and if the roots of A are $\alpha_1, \alpha_2, \cdots, \alpha_n$, then the roots of $T(A)$ have the form $\alpha_1^{r_1}\alpha_2^{r_2} \cdots \cdot \alpha_n^{r_n}$ where $\Sigma r_i = r$. The actual choice of the exponents r depends on the particular associated matrix in question but, if $\alpha_1^{r_1}\, \alpha_2^{r_2} \cdots \alpha_n^{r_n}$ is one root, all the distinct quantities obtained from it by permuting the α's are also roots.*

If the roots of A are arbitrary variables, then A is similar to a diagonal matrix $A_1 = \Sigma\alpha_i e_{ii}$. We can express $T(A_1)$ as a polynomial[6] in the α's, say

$$T(A_1) = \Sigma\alpha_1^{r_1}\alpha_2^{r_2} \cdots \alpha_n^{r_n} F_{r_1 r_2 \cdots r_n} \tag{52}$$

where the F's are constant matrices. If now $B = \Sigma\beta_i e_{ii}$ is a second variable diagonal matrix, the relation $T(A_1B) = T(A_1)T(B)$ gives as in (50)

$$\begin{aligned} F^2_{r_1 r_2 \cdots r_n} &= F_{r_1 r_2 \cdots r_n}, \\ F_{r_1 r_2 \cdots r_n} F_{s_1 s_2 \cdots s_n} &= 0 \ ((r_1, r_2, \cdots) \neq (s_1, s_2 \cdots)) \end{aligned} \tag{53}$$

and hence $T(A_1)$ can be expressed as a diagonal matrix with roots of the required form; these roots may of course be multiple since the rank of $F_{r_1 \cdots r_n}$ is not necessarily 1, the elementary divisors are, however, simple.

Since the associated matrices of similar matrices are similar, it follows that the roots of the characteristic equation of $T(A)$ are given by those terms in (52) for which $F_{r_1 r_2 \cdots r_n} \neq 0$; and, since this equation has coefficients which are polynomials in the coordinates of A, the roots of $T(A)$ remain in this form even when the roots of A are not necessarily all different.

The rest of the theorem follows from the fact that the trace of $T(A_1)$ equals that of $T(A)$ which is rational in the coordinates of A and is therefore symmetric in the α's.

THEOREM 11. *The value of the determinant of $T(A)$ is $|A|^{rm/n}$ and rm/n is an integer.*

For $T(A)T(\mathrm{adj}A) = T(|A|) = |A|^r$ and therefore $|T(A)|$ is a power of $|A|$, say $|A|^s$. But $T(A)$ is a matrix of order m whose coordinates are polynomials in the coordinates of A. Hence $sn = mr$ and rm/n is an integer.

5.15 **Transformable systems.** From a scalar point of view each of the associated matrices discussed in §§5.03–5.11 can be characterized by a set of scalar functions f_k $(k = 1, 2, \cdots, m)$ of one or more sets of variables $(\xi_i,$

[6] If we merely assume that $T(A_1)$ is a convergent series of the form (52), equation (53) still holds. It follows that there are only a finite number of terms in (52) since (53) shows that there is no linear relation among those $F_{r_1 \cdots r_n}$ which are not zero. Let F_i be the sum of those $F_{r_1 \cdots r_n}$ for which Σr_j has a fixed value ρ_i; then $T(\lambda) = \Sigma\lambda^{\rho_i}F_i$, and as before only one value of ρ_1 is admissible when $T(A)$ is irreducible.

$j = 1, 2, \cdots, n)$, $i = 1, 2, \cdots, r$, which have the following property: if the ξ's are subjected to a linear transformation

$$\xi'_{ij} = \sum_{s=1}^{n} a_{js}\xi_{is} \qquad (j = 1, 2, \cdots, n; i = 1, 2, \cdots, r)$$

and if f'_k is the result of replacing ξ_{ij} by ξ'_{ij} in f_k, then

$$f'_k = \sum_{s=1}^{m} \alpha_{ks} f_s$$

where the α's are functions of the a_{ij} and are independent of the ξ's. For instance, corresponding to $C_2(A)$ we have

$$f_i \equiv f_{pq} = \begin{vmatrix} \xi_{1p} & \xi_{1q} \\ \xi_{2p} & \xi_{2q} \end{vmatrix} \qquad (p, q = 1, 2, \cdots, n; p < q)$$

for which

$$\alpha_{ij} \equiv \alpha_{pq,\, rs} = \begin{vmatrix} a_{pr} & a_{qr} \\ a_{ps} & a_{qs} \end{vmatrix}.$$

We may, and will, always assume that there are no constant multipliers such that $\Sigma\lambda_i f_i = 0$. Such systems of functions were first considered by Sylvester; they are now generally called *transformable systems*.

If we put $T(A) = \| \alpha_{ij} \|$, we have immediately $T(AB) = T(A)T(B)$, and consequently there is an associated matrix corresponding to every transformable system. Conversely, there is a transformable system corresponding to an associated matrix. For if $X = \| \xi_{ij} \|$ is a variable matrix and c an arbitrary constant vector in the space of $T(A)$, then the coordinates of $T(X)c$ form a transformable system since $T(A)T(X)c = T(AX)c$ and c can be so determined that there is no constant vector b such that $SbT(X)c \equiv 0$.

The basis f_k $(k = 1, 2, \cdots, m)$ may of course be replaced by any basis which is equivalent in the sense of linear dependence, the result of such a change being to replace $T(A)$ by an equivalent associated matrix. If in particular there exists a basis

$$g_1, g_2, \cdots, g_{k_1}, h_1, h_2, \cdots, h_{k_2} \qquad (k_1 + k_2 = k)$$

such that the g's and the h's form separate transformable systems, then $T(A)$ is reducible; and conversely, if $T(A)$ is reducible, there always exists a basis of this kind.

5.16 Transformable linear sets. If we adopt the tensor point of view rather than the scalar one, an associated matrix is found to be connected with a linear set $\mathfrak{F}$ of constant tensors, derived from the fundamental units e_i, such that, when e_i is replaced by Ae_i $(i = 1, 2, \cdots, n)$ in the members of the basis of $\mathfrak{F}$, then the new tensors are linearly dependent on the old; in other words the set $\mathfrak{F}$ is invariant as a whole under any linear transformation A of the fundamental units. For instance, in the case of $C_2(A)$ cited above, $\mathfrak{F}$ is the linear set defined by

$$| e_i e_j | \qquad (i, j = 1, 2, \cdots, n; i < j).$$

We shall call a set which has this property a *transformable linear set.*

Let $u_1, u_2, \cdots, u_m$ be a transformable linear set of tensors of grade r and let u_i' be the tensor that results when e_j is replaced by Ae_j ($j = 1, 2, \cdots, n$) in u_i. Since the set is transformable, we have

$$u_i' = \sum_j \alpha_{ji} u_j = T(A) u_i \qquad (i = 1, 2, \cdots, m)$$

where the α_{ij} are homogeneous polynomials in the coordinates of A of degree r. If we employ a second transformation B, we then have

$$u_i'' = T(A)T(B)u_i, \qquad u_i'' = T(AB)u_i \qquad (i = 1, 2, \cdots, m)$$

and therefore $T(A)$ is an associated matrix.

We have now to show that there is a transformable linear set corresponding to every associated matrix. In doing this it is convenient to extend the notation Suv to the case where u and v are tensors of grade r. Let E_i ($i = 1, 2, \cdots, n^r$) be the unit tensors of grade r and

$$u = \Sigma \psi_i E_i, \; v = \Sigma \varphi_i E_i$$

any tensors of grade r; we then define Suv by

$$Suv = \left(\sum_1^{n^r} \psi_i \varphi_i \right) \bigg/ r!$$

where the numerical divisor is introduced solely in order not to disagree with the definition of §5.02.

Let $x_i = \Sigma \xi_{ij} e_j$ ($i = 1, 2, \cdots, r$) be a set of variable vectors and X_i ($i = 1, 2, \cdots, s$) the set of tensors of the form $x_1^{j_1} x_2^{j_2} \cdots x_r^{j_r}$ ($\Sigma j_i = r$); we can then put any product $\xi_{11}^{\beta_{11}} \xi_{12}^{\beta_{12}} \cdots \xi_{rn}^{\beta_{rn}}$ for which $\Sigma \beta_{ij} = r$ in the form kSE_iX_j, k being a numerical factor. This can be done in more than one way as a rule; in fact, if $\sum_j \beta_{ij} = \beta_i$, then

$$\xi_{11}^{\beta_{11}} \cdots \xi_{1n}^{\beta_{1n}} = \frac{1}{\beta_1!} S \, e_1^{\beta_{11}} \cdots e_n^{\beta_{1n}} \, x_1^{\beta_1}$$

and from the definition of Suv it is clear that the factors in $e_1^{\beta_{11}} \cdots e_n^{\beta_{1n}}$ can be permuted in any way without altering the value of the scalar. It follows that

$$\xi_{11}^{\beta_{11}} \cdots \xi_{1n}^{\beta_{1n}} = \frac{1}{\beta_{11}! \, \beta_{12}! \cdots \beta_{1n}!} S \begin{Bmatrix} e_1 & \cdots & e_n \\ \beta_{11} & \cdots & \beta_{1n} \end{Bmatrix} x_1^{\beta_1}$$

and repeating this process we get

$$k_1\xi_{11}^{\beta_{11}}\xi_{12}^{\beta_{12}} \cdots \xi_{rn}^{\beta_{rn}} = S\begin{Bmatrix} e_1 & \cdots & e_n \\ \beta_{11} & \cdots & \beta_{1n}\end{Bmatrix} \cdots \begin{Bmatrix} \cdots & e_n \\ \cdots & \beta_{rn}\end{Bmatrix} x_1^{\beta_1}x_2^{\beta_2} \cdots x_r^{\beta_r}$$

where k_1 is a numerical factor whose value is immaterial for our present purposes.

If f is any homogeneous polynomial in the variables ξ_{ij} of degree ρ, it can be expressed uniquely in the form

$$f = \Sigma\,\Sigma\varphi_{\beta_{11}\cdots\beta_{rn}} S\begin{Bmatrix} e_1 & \cdots & e_n \\ \beta_{11} & \cdots & \beta_{1n}\end{Bmatrix} \cdots \begin{Bmatrix} e_1 & \cdots & e_n \\ \beta_{r1} & \cdots & \beta_{rn}\end{Bmatrix} x_1^{\beta_1} \cdots x_r^{\beta_r}$$

where the inner summation extends over the partitions of β_i into $\beta_{i1}, \beta_{i2}, \cdots, \beta_{in}$ $(i = 1, 2, \cdots, r)$ and the outer over all values of $\beta_1, \beta_2, \cdots, \beta_r$ for which $\Sigma\beta_i = \rho$. We may therefore write

$$f = \sum_1^s SF_jX_j$$

where, as above, $X_j = x_1^{\beta_1}x_2^{\beta_2} \cdots x_r^{\beta_r}$ and

$$F_j \equiv F_{\beta_1\beta_2\cdots\beta_r} = \Sigma\varphi_{\beta_{11}\cdots\beta_{rn}}\begin{Bmatrix} e_1 & \cdots & e_n \\ \beta_{11} & \cdots & \beta_{1n}\end{Bmatrix} \cdots \begin{Bmatrix} e_1 & \cdots & e_n \\ \beta_{r1} & \cdots & \beta_{rn}\end{Bmatrix}.$$

The expression of f in this form is unique. In the first place, $F_j \not\equiv 0$ unless each $\varphi_{\beta_{11}\cdots\beta_{rn}}$ is zero, since the set of tensors of the form

$$\begin{Bmatrix} e_1 & \cdots & e_n \\ \beta_{11} & \cdots & \beta_{1n}\end{Bmatrix} \cdots \begin{Bmatrix} e_1 & \cdots & e_n \\ \beta_{r1} & \cdots & \beta_{rn}\end{Bmatrix} \quad (\Sigma\beta_{ij} = \rho)$$

are clearly linearly independent. Further, if $\Sigma SF_jX_j \equiv 0$, then each SF_jX_j is zero since each gives rise to terms of different type in the ξ_{ij}; and finally the form of F_j shows that $SF_jX_j = 0$ only if $F_j = 0$ since in

$$SF_jX_j = k_1\Sigma\varphi_{\beta_{11}\cdots\beta_{rn}}\xi_{11}^{\beta_{11}} \cdots \xi_{rn}^{\beta_{rn}}$$

each term of the summation is of different type in the ξ_{ij}.

Let (f_k) be a transformable system; we can now write uniquely

(54) $$f_k = \sum_j SF_{kj}X_j \quad (k = 1, 2, \cdots, m)$$

and we may set

$$F = \sum_1^{n^r} f_iE_i = \sum_{i,j} E_iSF_{ij}X_j$$

where $f_i \equiv 0$ when $i > m$. If we transform the x's by $A = \| a_{ij} \|$ and denote $\Pi_r(A)$ temporarily by Π, then X_j becomes ΠX_j and F is transformed into F^* where

(55) $$F^* = \sum_{i,j} E_iSF_{ij}\Pi X_j = \sum_{i,j} E_iS\,\Pi' F_{ij}\cdot X_j.$$

But the f's form a transformable system and hence by this transformation f_i becomes

$$f'_i = \sum_k \alpha_{ik} f_k$$

so that

$$F^* = \sum_{k,i} \alpha_{ik} f_k E_i = \sum_i E_i S \sum_k \alpha_{ik} \sum_j F_{kj} X_j. \tag{56}$$

Comparing (55) and (56) we have

$$\sum_j S \left[\sum_k \alpha_{ik} F_{kj} - \Pi' F_{ij} \right] X_j = 0 \tag{57}$$

and therefore, as was proved above, each of the terms of the summation is zero, that is,

$$\Pi' F_{ij} = \sum_k \alpha_{ik} F_{kj} \tag{58}$$

and therefore, if j is kept fixed, the linear set

$$(F_{1j}, F_{2j}, \cdots) \tag{59}$$

is transformable provided $F_{1j}, F_{2j}, \cdots$ are linearly independent.

If there is no j for which the set (59) is linearly independent we proceed as follows. Let $f_{ij} = SF_{ij}X_j$ so that

$$\begin{aligned} f_1 &= f_{11} + f_{12} + \cdots + f_{1s} \\ f_2 &= f_{21} + f_{22} + \cdots + f_{2s} \\ &\cdots\cdots\cdots\cdots\cdots\cdots \\ &\cdots\cdots\cdots\cdots\cdots\cdots \\ f_m &= f_{m1} + f_{m2} + \cdots + f_{ms}. \end{aligned} \tag{60}$$

If the removal of any column of this array leaves the new f_i so defined linearly independent, they form a transformable system which defines the same associated matrix as the original system; we shall therefore suppose that the removal of any column leads to linear relations among the rows, the coefficients of these relations being constants. Remove now the first column; then by non-singular constant combinations of the rows we can make certain of them, say the first m_1, equal 0, the remainder being linearly independent. On applying the same transformation to the rows of (60), which leaves it still a transformable system, we see that we may replace (60) by an array of the form

$$\begin{aligned} f_1 &= f_{11} \\ &\cdots\cdots\cdots \\ f_{m_1} &= f_{m_1 1} \\ f_{m_1+1} &= f_{m_1+1,1} + f_{m_1+1,2} + \cdots + f_{m_1+1,s} \\ &\cdots\cdots\cdots\cdots\cdots\cdots\cdots\cdots \\ f_m &= f_{m1} + f_{m2} + \cdots + f_{ms} \end{aligned} \tag{61}$$

where $f_{m_1+i} - f_{m_1+i,1}$ $(i = 1, 2, \cdots, m - m_1)$ are linearly independent. It follows that $f_1, \cdots, f_{m_1}$ are transformed among themselves and so form a transformable system. For these functions are transformed in the same way as $f_{11}, f_{21} \cdots, f_{m_11}$, and if the last $m - m_1$ rows of (61) were involved in the transformation, this would mean that $f_{11}, \cdots, f_{m_11}$, when transformed, would depend on $f_{m_1+1,i}$ etc., which is impossible owing to the linear independence of $f_{m_1+i} - f_{m_1+i,1}$ $(i = 1, 2, \cdots, m - m_1)$.

Corresponding to the first column of (61) we have tensors $F_{11}, F_{21}, \cdots, F_{m1}$ and we may suppose this basis so chosen that F_{i1} $(i = 1, 2, \cdots, p)$ are linearly independent and $F_{j1} = 0$ for $j > p$; and this can be done without disturbing the general form of (61). If $p = m$, we have a transformable system of the type we wish to obtain and we shall therefore assume that $p < m$. We may also suppose the basis so chosen that $S\bar{F}_{i1}F_{j1} = \delta_{ij}$ $(i, j = 1, 2, \cdots, p)$ as in Lemma 2, §1.09. It follows from what we have proved above that $F_{11}, F_{21}, \cdots, F_{m_11}$ is a transformable set.

Let A be a real matrix, the corresponding transformation of the F's being, as in (58),

$$F^*_{i1} = \sum_j \alpha_{ij}F_{j1} = \Pi'F_{i1}, \quad (i = 1, 2, \cdots, p); \tag{62}$$

we then have

$$\bar{F}^*_{i1} = \sum_j \bar{\alpha}_{ij}\bar{F}_{j1} = \Pi'(A)\bar{F}_{i1} \tag{63}$$

so that the $\bar{F}_{i1}$ also forms a transformable set. Since $F_{11}, \cdots, F_{m_11}$ form a transformable set, α_{ij} and $\bar{\alpha}_{ij}$ are 0 when $i > m_1$ and $j \le m_1$ no matter what matrix A is. Now

$$\alpha_{ij} = S\bar{F}_{j1}F^*_{i1} = S\bar{F}_{j1}\Pi'(A)F_{i1} = S\Pi(A)\bar{F}_{j_1}F_{i_1} = S\Pi'(A')\bar{F}_{j_1}F_{i_1}$$

which equals 0 for $i \le m_1, j > m_1$ since by (63) $\Pi'(A')\bar{F}_{j1}$ is derived from $\bar{F}_{j1}$ by the transformation A' on the x's and for $j \le m_1$ is therefore linearly dependent on $\bar{F}_{j1}$ $(j = 1, 2, \cdots, m_1)$. Hence the last $m - m_1$ rows in (61) also form a transformable system, which is only possible if the system $f_1, f_2, \cdots, f_m$ is reducible. If $T(A)$ is irreducible, the corresponding transformable system is irreducible and it follows now that there also corresponds to it an irreducible transformable set of tensors.

5.17 We have now shown that to every associated matrix $T(A)$ of index r and order m there corresponds a transformable linear set of constant tensors $F_1, F_2, \cdots, F_m$ of grade r whose law of transformation is given by (62). Also since $\Pi'(A) = \Pi(A')$, we have

$$\Pi F_i = \Sigma\alpha'_{ik}F_k, \quad \Pi\bar{F}_i = \Sigma\bar{\alpha}'_{ik}\bar{F}_k \tag{64}$$

where $T(A') = \| \alpha'_{ij} \|$.

Since $F_1, F_2, \cdots, F_m$ are linearly independent, we can find a supplement to this set in the set of all tensors of grade r, say

$$G_1, G_2, \cdots, G_\mu \qquad (\mu = n^r - m)$$

such that

(65) $$S\bar{F}_i G_j = 0.$$

It is convenient also to choose bases for both sets such that

(65′) $$S\bar{F}_i F_j = \delta_{ij} = S\bar{G}_i G_j.$$

Since the two sets together form a basis for the space of Π, we can set

$$\Pi' G_j = \Sigma\beta_{kj}F_k + \Sigma\gamma_{kj}G_k$$

and this gives

$$\beta_{ij} = S\bar{F}_i\Pi' G_j = SG_j\Pi\bar{F}_i$$

which is 0 from (64) and (65), hence the G's are transformed among themselves by Π'. This means, however, that Π' is reducible, and when it is expressed in terms of the basis $(F_1, \cdots, F_m, G_1, \cdots, G_\mu)$, the part corresponding to $(F_1, \cdots, F_m)$ has the form $\| \alpha_{ij} \|$ and is therefore similar to $T(A)$. Hence:

THEOREM 12. *Every irreducible associated matrix $T(A)$ of index r is equivalent to an irreducible part of $\Pi_r(A)$, and conversely.*

5.18 Irreducible transformable sets. If F is a member of a transformable linear set $\mathfrak{F} = (F_1, F_2, \cdots, F_m)$, the total set of tensors derived from F by all linear transformations of the fundamental units clearly form a transformable linear set which is contained in $\mathfrak{F}$, say $\mathfrak{F}_1$; and we may suppose the basis of $\mathfrak{F}$ so chosen that $\mathfrak{F}_1 = (F_1, F_2, \cdots, F_k)$ and $S\bar{F}_iF_j = \delta_{ij}$ $(i, j = 1, 2, \cdots, m)$. Let G be an element of $(F_{k+1}, \cdots, F_m)$ and G' a transform of G so that

$$G' = \sum_{i=1}^{m} \gamma_i F_i.$$

Then $S\bar{F}_iG' = \gamma_i$. But $S\bar{F}_iG' = S\bar{F}'_iG$, where F'_i is the transform of F_i obtained by the transverse of the transformation which produced G' from G so that $\bar{F}'_i$ is in $\mathfrak{F}_1$ for $i \leq k$. Hence $\gamma_i = 0$ for $i = 1, 2, \cdots, k$, that is, $(F_{k+1}, \cdots, F_m)$ is also a transformable set; and so, when the original set is irreducible, we must have $\mathfrak{F}_1 = \mathfrak{F}$. If we say that F generates $\mathfrak{F}$, this result may be stated as follows.

LEMMA 5. *An irreducible transformable linear set is generated by any one of its members.*

We may choose F so that it is homogeneous in each e_i; for if we replace, say, e_1 by λe_1, then F has the form $\Sigma\lambda^k H_k$ and by the same argument as in §5.13, any H_k which is not 0 is homogeneous in e_1 and belongs to $\mathfrak{F}$. A repetition of this argument shows that we may choose F to be homogeneous in each of the fundamental units which occur in it. If r is the grade of F, we may assume that F depends on $e_1, e_2, \cdots, e_s$, and, if $k_1, k_2, \cdots, k_s$ are the corresponding degrees of homogeneity, then $\Sigma k_i = r$ and, when convenient, we may arrange the notation so that $k_1 \geq k_2 \geq \cdots \geq k_s$.

If we now replace e_1 in F by $e_1 + \lambda e_i$ $(i > s)$, the coefficient H of λ is not 0, since $i > s$, and H becomes $k_1 F$ when e_1 is replaced by e_1; it therefore forms a generator of $\mathfrak{F}$ in which the degree of e_1 is one less than before. It follows that, when $r \leq n$, we may choose a generator which is linear and homogeneous in r units $e_1, e_2, \cdots, e_r$. It is also readily shown that such a tensor defines an irreducible transformable linear set if, and only if, it forms an irreducible set when the transformations of the units are restricted to permuting the first r e's among themselves. Further, since the choice of fundamental units is arbitrary, we may replace them by variable vectors $x_1, x_2, \cdots, x_r$. For instance, the transformable sets associated with Π_r, P_r and C_r are $x_1x_2 \cdots x_r$, $\{x_1x_2 \cdots x_r\}$ and $|x_1x_2 \cdots x_r|$, respectively, and of these the first is reducible and the other two irreducible.

5.19 It is not difficult to calculate directly the irreducible transformable sets for small values of r by the aid of the results of the preceding paragraph. If we denote $x_1, x_2, \cdots$ by $1, 2, \cdots$, the following are generators for $r = 2, 3$.

	generator		order
		$r = 2$	
2.1	$\{12\}$		$n(n + 1)/2$
2.2	$\lvert 12 \rvert$		$n(n - 1)/2$
		$r = 3$	
3.1	$\{123\}$		$n(n + 1)(n + 2)/6$
3.2	$\lvert 1 \lvert 23 \rvert \rvert$		$n(n^2 - 1)/3$
3.3	$\lvert 1\{23\} \rvert$		$n(n^2 - 1)/3$
3.4	$\lvert 123 \rvert$		$n(n - 1)(n - 2)/6.$

This method of determining the generators directly is tedious and the following method is preferable.[7] Any generator has the form

$$w_1 = \Sigma \omega_{i_1 i_2 \cdots i_r} x_{i_1} x_{i_2} \cdots x_{i_r}$$

and if $q_{i_1 \cdots i_r}$ denotes the substitution $\begin{pmatrix} 1, & 2, & \cdots & r \\ i_1 & i_2 & \cdots & i_r \end{pmatrix}$, we may write

$$\begin{aligned} w_1 &= \Sigma \omega_{i_1 i_2 \cdots i_r} q_{i_1 i_2 \cdots i_r} x_1 x_2 \cdots x_r \\ &= q_1(x_1 x_2 \cdots x_r) \end{aligned}$$

where q_1 may be regarded (see chap. 10) as an element of the algebra S whose units are the operators q of the symmetric group on r letters. Now w_1 generates a transformable set and hence, if $w_i = q_i(x_1 \cdots x_r)$ $(i = 1, 2, \cdots)$ is a

[7] Fuller details of the actual determination of the generators will be found in Weyl: *Gruppentheorie und Quantentheorie*, 2 ed. chap. 5.

basis of the set, and Q is the set of elements $q_1, q_2, \cdots$ in S, then the set of elements $Qq = (q_1q, q_2q, \cdots)$ must be the same as the set Q, that is, in the terminology of chapter 10, Q is a semi-invariant subalgebra of S; conversely any such semi-invariant subalgebra gives rise to a transformable set and this set is irreducible if the semi-invariant subalgebra is minimal, that is, is contained in no other such subalgebra.

It follows now from the form derived for a group algebra such as S that we get all independent generators as follows. In the first place the operators of S can be divided into sets[8] S_k $(k = 1, 2, \cdots, t)$ such that (i) the product of an element of S_k into an element of S_j $(k \neq j)$ is zero; (ii) in the field of complex numbers a basis for each S_k can be chosen which gives the algebra of matrices of order n_k^2; and in an arbitrary field S is the direct product of a matric algebra and a division algebra; (iii) there exists a set of elements $u_{k1}, u_{k2}, \cdots, u_{k\nu_k}$ in S_k such that $\sum_i u_{ki}$ is the identity of S_k and $u_{ki}^2 = u_{ki} \neq 0$, $u_{ki}u_{kj} = 0$ $(i \neq j)$ and such that the set of elements $u_{ki}S_ku_{ki}$ is a division algebra, which in the case of the complex field contains only one independent element; (iv) the elements of S_k can be divided into ν_k sets $u_{ki}S_k$ $(i = 1, 2, \cdots)$ each of which is a minimal semi-invariant subalgebra of S and therefore corresponds to an irreducible transformable set.

[8] It is shown in the theory of groups that t equals the number of partitions of r.

CHAPTER VI

SYMMETRIC, SKEW, AND HERMITIAN MATRICES

6.01 Hermitian matrices. If we denote by $\bar{x}$ the matrix which is derived from x by replacing each coordinate by its conjugate imaginary, then x is called a *hermitian* matrix if

$$\bar{x} = x'. \tag{1}$$

We may always set $x = x_1 + ix_2$ where x_1 and x_2 are real and (1) shows that, when x is hermitian,

$$x_1' = x_1, \; x_2' = -x_2, \tag{2}$$

so that the theory of real symmetric and real skew matrices is contained in that of the hermitian matrix. The following are a few properties which follow immediately from the definition; their proof is left to the reader.

If x and y are hermitian and a is arbitrary, then

$$x + y, \;\; \bar{x}, \;\; x', \;\; ax\bar{a}', \;\; xy + yx, \;\; i(xy - yx),$$

are all hermitian.

Any matrix x can be expressed uniquely in the form $a + ib$ where $2a = x + \bar{x}'$, $2b = -i(x - \bar{x}')$ are hermitian.

The product of two commutative hermitian matrices is hermitian. In particular, any integral power of a hermitian matrix x is hermitian; and, if $g(\lambda)$ is a scalar polynomial with real coefficients, $g(x)$ is hermitian.

THEOREM 1. *If $a, b, c, \cdots$ are hermitian matrices such that $a^2 + b^2 + c^2 + \cdots = 0$, then $a, b, c, \cdots$ are all 0.*

If $\Sigma a^2 = 0$, its trace is 0; but $\Sigma a^2 = \Sigma a\bar{a}'$ and the trace of the latter is the sum of the squares of the absolute values of the coordinates of $a, b, \cdots$; hence each of these coordinates is 0.

THEOREM 2. *The roots of a hermitian matrix are real and its elementary divisors are simple.*

Let x be a hermitian matrix and $g(\lambda)$ its reduced characteristic function. Since $g(x) = 0$, we have $0 = \bar{g}(\bar{x}) = \bar{g}(x')$ and, since x and x' have the same reduced characteristic function, it follows that $g(\lambda) \equiv \bar{g}(\lambda)$, that is, the coefficients of g are real. Suppose that $\xi_1 = \alpha + i\beta$ ($\beta \neq 0$) is a root of $g(\lambda)$; then $\xi_2 = \alpha - i\beta \neq \xi_1$ is also a root, and we may set

$$g(\lambda) = (\lambda - \xi_1)(g_1(\lambda) + ig_2(\lambda)) = (\lambda - \xi_2)(g_1(\lambda) - ig_2(\lambda)) \tag{3}$$

where g_1, g_2 are real polynomials of lower degree than g, neither of which is identically 0 since g is real and ξ_1 complex. Now

$$[g_1(x)]^2 + [g_2(x)]^2 = [g_1(x) + ig_2(x)][g_1(x) - ig_2(x)]$$

and this product is 0 since from (3) $\lambda - \xi_1$ is a factor of $g_1(\lambda) - ig_2(\lambda)$ and $(\lambda - \xi_1)(g_1(\lambda) + ig_2(\lambda)) = g(\lambda)$. But, since the coefficients of g_1 and g_2 are real, the matrices $g_1(x)$, $g_2(x)$ are hermitian and, seeing that the sum of their squares is 0, they both vanish by Theorem 1. This is however impossible since $g_1(\lambda)$ is of lower degree than the reduced characteristic function of x. Hence x cannot have a complex root.

To prove that the elementary divisors are simple it is only necessary to show that $g(\lambda)$ has no multiple root. Let

$$g(\lambda) = (\lambda - \xi)^r h(\lambda), \quad h(\xi) \neq 0.$$

If $r > 1$, set $g_1(\lambda) = (\lambda - \xi)^{r-1}h(\lambda)$; then $[g_1(\lambda)]^2$ has $g(\lambda)$ as a factor so that the square of the hermitian matrix $g_1(x)$ is 0. Hence by Theorem 1, $g_1(x)$ is itself 0, which is impossible since the degree of g_1 is less than that of g. It follows that r cannot be greater than 1, which completes the proof of the theorem.

Since the elementary divisors are simple, the canonical form of x is a diagonal matrix. Suppose that $n - r$ roots are 0 and that the remaining roots are $\xi_1, \xi_2, \cdots, \xi_r$; these are of course not necessarily all different. The canonical form is then

$$\begin{matrix} \xi_1 & & & & & & \\ & \xi_2 & & & & & \\ & & \ddots & & & & \\ & & & \xi_r & & & \\ & & & & 0 & & \\ & & & & & \ddots & \\ & & & & & & 0. \end{matrix}$$

The following theorem is contained in the above results.

THEOREM 3. *A hermitian matrix of rank r has exactly $n - r$ zero roots.*

It also follows immediately that the characteristic equation of x has the form

$$x^n - a_1x^{n-1} + \cdots + (-1)^r a_r x^{n-r} = 0 \quad (a_r \neq 0)$$

where a_i is the elementary symmetric function of the ξ's of degree i. Since a_r is the sum of the principal minors of x of order r, we have

THEOREM 4. *In a hermitian matrix of rank r at least one principal minor of order r is not 0.*

In view of the opening paragraph of this section Theorems 1–4 apply also to real symmetric matrices; they apply also to real skew matrices except that Theorem 2 must be modified to state that the roots are pure imaginaries.

6.02 The invariant vectors of a hermitian matrix. Let H be a hermitian matrix, α_1, α_2 two different roots, and a_1, a_2 the corresponding invariant vectors so chosen that $Sa_i\bar{a}_i = 1$; then, since $Ha_1 = \alpha_1 a_1$, $\bar{H}\bar{a}_1 = \alpha_1\bar{a}_1$, we have

$$\alpha_1 Sa_2\bar{a}_1 = Sa_2\bar{H}\bar{a}_1 = S\bar{H}'a_2\bar{a}_1 = \alpha_2 Sa_2\bar{a}_1$$

and, since $\alpha_1 \neq \alpha_2$, we must have $Sa_2\bar{a}_1 = 0$. Again, if α is a repeated root of order s and $a_1, a_2, \cdots, a_s$ a corresponding set of invariant vectors we may choose these vectors (cf. §1.09) so that $Sa_i\bar{a}_j = \delta_{ij}$. The invariant vectors may therefore be so chosen that they form a unitary set and

(4) $$H = \Sigma\alpha_i a_i S\bar{a}_i.$$

If U is the matrix defined by

(5) $$Ue_i = a_i \quad (i = 1, 2, \cdots, n),$$

then

(6) $$U\bar{U}' = 1,$$

so that U is unitary, and if A is the diagonal matrix $\sum_1^r \alpha_i e_i S e_i$, then

(7) $$H = UA\bar{U}' = UAU^{-1}.$$

We may therefore say:

THEOREM 5. *A hermitian matrix can be transformed to its canonical form by a unitary matrix.*

If H is a real symmetric matrix, the roots and invariant vectors are real, and hence U is a real orthogonal matrix. Hence

THEOREM 6. *A real symmetric matrix can be transformed to its canonical form by a real orthogonal matrix.*

If T is a real skew matrix, $h = iT$ is hermitian. The non-zero roots of T are therefore pure imaginaries and occur in pairs of opposite sign. The invariant vectors corresponding to the zero roots are real and hence by the proof just given they may be taken orthogonal to each other and to each of the other invariant vectors. Hence, if the rank of T is r, we can find a real orthogonal matrix which transforms it into a form in which the last $n - r$ rows and columns are zero.

Let $i\alpha$ be a root of T which is not 0 and $a = b + ic$ a corresponding invariant vector; then $Ta = i\alpha a$ so that

$$Tb = -\alpha c, \quad Tc = \alpha b.$$

Hence

$$-\alpha Sc^2 = ScTb = -SbTc = -\alpha Sb^2, \quad -\alpha Sbc = SbTb = 0,$$

which gives

$$Sb^2 = Sc^2, \quad Sbc = 0.$$

We can then choose a so that $Sb^2 = Sc^2 = 1$ and can therefore find a real orthogonal matrix which transforms T into

$$\begin{matrix} 0 & \alpha_1 & & & \\ -\alpha_1 & 0 & & & \\ & & 0 & \alpha_2 & \\ & & -\alpha_2 & 0 & \\ & & & & \ddots \end{matrix} = \Sigma\alpha_j(e_{2j-1}Se_{2j} - e_{2j}Se_{2j-1}). \tag{8}$$

We have therefore the following theorem.

THEOREM 7. *If T is a real skew matrix, its non-zero roots are pure imaginaries and occur in pairs of opposite sign; its rank is even; and it can be transformed into the form* (8) *by a real orthogonal matrix.*

6.03 **Unitary and orthogonal matrices.** The following properties of a unitary matrix follow immediately from its definition by equation (6).

The product of two unitary matrices is unitary.

The transform of a hermitian matrix by a unitary matrix is hermitian.

The transform of a unitary matrix by a unitary matrix is unitary.

If H_1 and H_2 are hermitian, a short calculation shows that

$$U_1 = \frac{1 - iH_1}{1 + iH_1}, \qquad U_2 = \frac{iH_2 - 1}{iH_2 + 1} \tag{9}$$

are unitary (the inverses used exist since a hermitian matrix has only real roots). Solving (9) for H_1 and H_2 on the assumption that the requisite inverses exist we get

$$H_1 = \frac{i(U_1 - 1)}{U_1 + 1}, \qquad H_2 = \frac{i(U_2 + 1)}{U_2 - 1}.$$

These are hermitian when U_1 and U_2 are unitary, and therefore any unitary matrix which has no root equal to -1 can be put in the first form while the second can be used when U has no root equal to 1.

THEOREM 8. *The absolute value of each root of a unitary matrix equals* 1.

Let $\alpha + i\beta$ be a root and $a + ib$ a corresponding invariant vector; then

$$U(a + ib) = (\alpha + i\beta)(a + ib), \overline{U}(a - ib) = (\alpha - i\beta)(a - ib).$$

Hence

$$Sa^2 + Sb^2 = S(a + ib)(a - ib) = S(a + ib)U'\overline{U}(a - ib) = SU(a + ib)\overline{U}(a - ib)$$
$$= (\alpha^2 + \beta^2)S(a + ib)(a - ib) = (\alpha^2 + \beta^2)S(a^2 + b^2),$$

so that $\alpha^2 + \beta^2 = 1$.

Corollary.

$$U^{-1}(a + ib) = (\alpha - i\beta)(a + ib),$$
$$U'(a - ib) = \overline{U}^{-1}(a - ib) = (\alpha + i\beta)(a - ib). \tag{10}$$

Theorem 9. *The elementary divisors of a unitary matrix are simple.*

For, if we have
$U(a_1 + ib_1) = (\alpha + i\beta)(a_1 + ib_1)$, $U(a_2 + ib_2) = (\alpha + i\beta)(a_2 + ib_2) + (a_1 + ib_1)$,
then from (10)

$$(\alpha + i\beta)S(a_1 - ib_1)(a_2 + ib_2) = SU'(a_1 - ib_1)(a_2 + ib_2) = S(a_1 - ib_1)U(a_2 + ib_2)$$
$$= (\alpha + i\beta)S(a_1 - ib_1)(a_2 + ib_2) + S(a_1 - ib_1)(a_1 + ib_1)$$

which is impossible since $S(a_1 - ib_1)(a_1 + ib_1) = Sa_1^2 + Sb_1^2 \neq 0$.

The results of this section apply immediately to real[1] orthogonal matrices; it is however convenient to repeat (9).

Theorem 10. *If U is a real orthogonal matrix, it can be expressed in the form $(1 + T)/(1 - T)$ if it has no root equal to 1 and in the form $(T - 1)/(T + 1)$ if it has no root equal to -1, the matrix T being a real skew matrix in both cases; and any real matrix of this form which is not infinite, is a real orthogonal matrix.*

6.04 Hermitian and quasi-hermitian forms. Let H be a hermitian matrix and $x = u + iv$ a vector of which u and iv are the real and imaginary parts; then the bilinear form $f = S\bar{x}Hx$ is called a *hermitian form.* Such a form is real since

$$\bar{f} = Sx\bar{H}\bar{x} = SxH'\bar{x} = S\bar{x}Hx = f.$$

In particular, if x and H are real, f is a real quadratic form and, if $H = iT$ is a pure imaginary, T is skew and $f = 0$.

If we express H in terms of its invariant vectors, say $H = \Sigma\alpha_i a_i S\bar{a}_i$ and then put $x = \Sigma\xi_i a_i$, the form f becomes $f = \Sigma\alpha_i\xi_i\bar{\xi}_i$. This shows that, if all the roots of H are positive, the value of f is positive for all values of x; H and f are then said to be *positive definite.* Similarly if all the roots are negative, H and f are *negative definite.* If some roots are 0 so that f vanishes for some value of $x \neq 0$, we say that H and f are *semi-definite,* positive or negative as the case may be. It follows immediately that, when H is semi-definite, $S\bar{x}Hx$ can only vanish if $Hx = 0$.

[1] The first part of the theorem applies also to complex orthogonal matrices.

THEOREM 11. *If H and K are hermitian and H is definite, the elementary divisors of $H\lambda - K$ are real and simple.*

Since $H\lambda - K$ and $-(H\lambda - K)$ are equivalent, we may suppose that H is positive definite. Its roots are then positive so that

$$H^{\frac{1}{2}} = \Sigma \alpha_i^{\frac{1}{2}} a_i S \bar{a}_i$$

has real roots and hence is also hermitian so that $H^{-\frac{1}{2}}KH^{-\frac{1}{2}}$ is hermitian. But

$$H\lambda - K = H^{\frac{1}{2}}(\lambda - H^{-\frac{1}{2}}KH^{-\frac{1}{2}})H^{\frac{1}{2}}$$

so that $H\lambda - K$ is equivalent to $\lambda - H^{-\frac{1}{2}}KH^{-\frac{1}{2}}$ which has real and simple elementary divisors by Theorem 2.

In order to include the theory of complex symmetric matrices we shall now define a type of matrix somewhat more general than the hermitian matrix and closely connected with it. If $A = A(\lambda)$ is a matrix whose coefficients are analytic functions of a scalar variable λ, we shall call it *quasi-hermitian* if

$$A'(\lambda) = A(-\lambda). \tag{11}$$

For convenience we shall set $A''(\lambda)$ for $A(-\lambda)$ with a similar convention for vector functions.

If $A = B + \lambda C$, B and C being functions of λ^2, then $A'' = B - \lambda C$ so that, if A is quasi-hermitian, B is symmetric and C skew just as in the case of a hermitian matrix except that now B and C are not necessarily real. If A is any matrix,

$$2P' \equiv A' + A'' = 2P'', \qquad 2Q' \equiv (A' - A'')/\lambda = 2Q''$$

so that any matrix can be expressed in the form $P + \lambda Q$ where P and Q are quasi-hermitian.

If $x = u + \lambda v$, where u and v are vectors which are functions of λ^2 and if A is quasi-hermitian, then

$$f(\lambda) = Sx''Ax = f(-\lambda) \tag{12}$$

is called a *quasi-hermitian form*. Again, if $|1 + \lambda A| \not\equiv 0$, and we set $Q = (1 - \lambda A)/(1 + \lambda A)$, then

$$Q' = \frac{(1 - \lambda A')}{1 + \lambda A'} = \frac{1 - \lambda A''}{1 + \lambda A''} = (Q'')^{-1}$$

so that

$$Q'Q'' = 1. \tag{13}$$

We shall call such a matrix *quasi-orthogonal*.

6.05 Reduction of a quasi-hermitian form to the sum of squares. We have seen in §5.06 that any matrix A of rank r can be expressed in the form

$$A = \sum_{s=1}^{r} A_s y_s \frac{SA_s' x_s}{Sx_s A_s y_s} \tag{14}$$

where

$$A_{s+1} = A_s - A_s y_s \frac{SA'_s x_s}{Sx_s A_s y_s}, \quad A_1 = A, \quad Sx_s A_s y_s \neq 0,$$

and the null space of A_{s+1} is obtained by adding $(y_1, y_2, \cdots, y_s)$ to the null space of A and the null space of A'_{s+1} by adding $(x_1, x_2, \cdots, x_s)$ to that of A'.

Suppose now that A is quasi-hermitian and replace y_s, x_s by z_s, z''_s and set $z_s = u_s + \lambda v_s$, $A_s = B_s + \lambda C_s$ so that

$$Sz''_s A_s z_s = Su_s B_s u_s + \lambda^2(2Su_s C_s v_s - Sv_s B_s v_s)$$

and, so long as A_s is not 0, we can clearly choose z_s so that $Sz''_s A_s z_s \not\equiv 0$. Each matrix A_s is then quasi-hermitian since $A'_s = A''_s$, and

(15)
$$A = \sum_1^r \frac{A_s z_s S A''_s z''_s}{Sz''_s A_s z_s}.$$

If x is an arbitrary vector and

$$f = f(\lambda^2) = SxAx'', \varphi_s(\lambda) = SxA_s z_s = \psi_s(\lambda^2) + \lambda\chi_s(\lambda^2)$$

then ψ_s and χ_s are linear functions of the coordinates of x which are linearly independent and

(16)
$$f = \sum_s \frac{SxA_s z_s S A''_s z''_s \cdot x''}{Sz''_s A_s z_s} = \sum \frac{\varphi_s(\lambda)\varphi''_s(\lambda)}{Sz''_s A_s z_s} = \sum \frac{\psi_s^2(\lambda^2) - \lambda^2\chi_s^2(\lambda^2)}{Sz''_s A_s z_s}$$

which is the required expression for $f(\lambda^2)$ in terms of squares.

If A is hermitian, then $\lambda = i$ and ψ_s, χ_s, $Sz''_s A_s z_s = S\bar{z}_s A_s z_s$ are real and, if $S\bar{z}_s A_s z_s = \alpha_s^{-1}$, (16) becomes

(17)
$$f = \sum_1^r \alpha_s \varphi_s \bar{\varphi}_s.$$

If $\lambda = 0$, then A is symmetric and

(18)
$$f = SxAx = \sum_1^r \alpha_s \varphi_s^2$$

where the terms are all real if A is real.

In terms of the matrices themselves these results may be expressed as follows.

Theorem 12. *If A is a hermitian matrix of rank r, there exist an infinity of sets of vectors a_s and real constants α_s such that*

(19)
$$A = \sum_1^r \alpha_s a_s S\bar{a}_s;$$

and, if A is symmetric, there exists an infinity of sets of vectors a_s and constants α_s such that

$$A = \sum_{1}^{r} \alpha_s a_s S a_s \tag{20}$$

a_s and α_s being real if A is real.

If π of the α's in (19) are positive and ν are negative, the difference $\sigma = \pi - \nu$ is called the *signature* of A. A given hermitian matrix may be brought to the form (19) in a great variety of ways but, as we shall now show, the signature is the same no matter how the reduction is carried out. Let K_1 be the sum of the terms in (19) for which α_s is positive and $-K_2$ the sum of the terms for which it is negative so that $A = K_1 - K_2$; the matrices K_1 and K_2 are positive semi-definite and, if k_1 and k_2 are their ranks, we have $r = k_1 + k_2$. Suppose that by a different method of reduction we get $A = M_1 - M_2$ where M_1 and M_2 are positive semi-definite matrices of ranks m_1 and m_2 and $m_1 + m_2 = r$; and suppose, if possible, that $k_2 < m_2$. The orders of the null spaces of K_2 and M_1 relative to the right ground of A are $r - k_2$ and $r - m_1 = m_2$ and, since $r - k_2 + m_2 > r$, there is at least one vector x in the ground of A which is common to both these null spaces, that is,

$$Ax = K_1 x = -M_2 x \neq 0,$$

and hence $S\bar{x}K_1x = -S\bar{x}M_2x$. But both K_1 and M_2 are positive semi-definite; hence we must have $S\bar{x}K_1x = 0$ which by §6.04 entails $K_1x = 0$. We have therefore arrived at a contradiction and so must have $k_2 = m_2$ which is only possible when the signature is the same in both cases.

In the case of a skew matrix the reduction given by (16) is not convenient and it is better to modify it as follows. Let $A' = -A$ and set

$$A_{s+1} = A_s + \frac{A_s y_s S A_s x_s}{S x_s A_s y_s} - \frac{A_s x_s S A_s y_s}{S x_s A_s y_s}, \tag{21}$$
$$A_1 = A, \qquad S x_s A_s y_s \neq 0.$$

So long as $A_s \neq 0$, the condition $Sx_sA_sy_s \neq 0$ can always be satisfied by a suitable choice of x_s and y_s and it is easily proved as in §5.06 that the null space of A_{s+1} is obtained from that of A_s by adding x_s, y_s; also A_s is skew so that we must necessarily have $x_s \neq y_s$. It follows that the rank of A is even and

$$A = \sum_{1}^{r/2} \frac{A_s x_s S A_s y_s - A_s y_s S A_s x_s}{S x_s A_s y_s} = \Sigma \alpha_s (a_{2s-1} S a_{2s} - a_{2s} S a_{2s-1}) \tag{22}$$

where each term in the summation is a skew matrix of rank 2 and

$$\alpha_s^{-1} = S x_s A_s y_s, \qquad a_{2s-1} = A_s x_s, \qquad a_{2s} = A_s y_s.$$

This form corresponds to the one given in Theorem 12 for symmetric matrices.

If we put

$$T = \sum_1^{r/2} (e_{2s-1}Se_{2s} - e_{2s}Se_{2s-1}) = -T'$$

(23)
$$R = \sum_1^{r/2} \alpha_s(e_{2s-1}Se_{2s} - e_{2s}Se_{2s-1}) = -R'$$

$$P = \sum_1^{r/2} (\alpha_s a_{2s}Se_{2s} + a_{2s-1}Se_{2s-1}), \qquad Q = \sum_1^{r/2} (a_{2s}Se_{2s} + a_{2s-1}Se_{2s-1})$$

then (22) may be put in the form $A = PTP' = QRQ'$. When $r = n$, the determinant of T equals 1 and therefore $|A| = |P|^2$. The following theorem summarizes these results.

Theorem 13. *If A is a skew matrix of rank r, then* (i) *r is even;* (ii) *A can be expressed by rational processes in the form*

(24)
$$A = \sum_1^{r/2} \alpha_s(a_{2s-1}Sa_{2s} - a_{2s}Sa_{2s-1}) = PTP' = QRQ'$$

where P, Q, R and T are given by (23)*;* (iii) *if $r = n$, the determinant of A is a perfect square, namely $|P|^2$;* (iv) *if x and y are any vectors and $w = \Sigma\alpha_s |a_{2s-1}a_{2s}|$, then*

(25)
$$SxAy = S|xy|w.$$

The following theorem contains several known properties of hermitian matrices.

Theorem 14. *If $T(A)$ is an associated matrix for which $T'(A) = T(A')$, then, when A is quasi-hermitian, $T(A)$ is also quasi-hermitian.*

For $A' = A''$ gives $T'(A) = T(A') = T(A'') = T''(A)$.

Particular cases of interest are: If A is hermitian, $T(A)$ is hermitian. If $T(\mu A) = \mu^s T(A)$ and A is skew, then $T(A)$ is skew if s is odd, symmetric if s is even.

6.06 The Kronecker method of reduction. Let $A = \sum_1^r x_i S y_i$ be a quasi-hermitian matrix of rank r; then

(26)
$$\Sigma y_i S x_i = A' = A'' = \Sigma x_i'' S y_i'',$$

from which it follows that y_i is linearly dependent on x_1'', x_2'', $\cdots$, x_r'', say

$$y_i = \sum_{j=1}^r q_{ij} x_j'', \qquad |q_{ij}| \neq 0, \qquad (i = 1, 2, \cdots, r).$$

Using this value of y_i we have

$$A = \Sigma q_{ij} x_i S x''_j, \qquad A' = \Sigma q_{ji} x''_i S x_j, \qquad A'' = \Sigma q''_{ij} x''_i S x_j$$

and therefore

(27) $$q_{ij} = q''_{ji}.$$

Further, since $| q_{ij} | \neq 0$, we can find s_{ij} $(i, j = 1, 2, \cdots, r)$ so that

$$\sum_j q_{ij} s_{jk} = \delta_{ik}$$

and then (27) gives $s_{ij} = s''_{ji}$.

Let $x_1, \cdots, x_r, x_{r+1}, \cdots, x_n$ be a basis and $z_1, z_2, \cdots, z_n$ the reciprocal basis. Then, if $w_i = \sum_1^r s_{ji} z''_j$, the basis reciprocal to $y''_1, \cdots, y''_r, x_{r+1}, \cdots, x_n$ is $w''_1, \cdots, w''_r, z_{r+1}, \cdots, z_n$. Hence

$$P = \sum_1^r w_i S z_i = \Sigma s_{ij} z''_i S z_j$$

is quasi-hermitian. Further, if $u = \sum_1^r \xi_i x_i$, then $Su''Pu = \Sigma \xi''_i \xi_j s_{ij}$; and we can choose u so that this form is not 0. We also have

$$AP = \sum_1^r x_i S y_i \sum_1^r w_j S z_j = \sum_1^r x_i S z_i,$$

whence $APu = u$.

Let

(28) $$A_{s+1} = A_s - \frac{u_s S u''_s}{S u''_s P_s u_s}, \qquad A_1 = A, \qquad P_1 = P,$$

where P_s is formed from A_s in the same way as P is from A and u_s is a vector of the left ground of A_s such that $Su''_s P_s u_s \neq 0$; also, as above, $A_s P_s u = u$ for any vector u in the left ground of A_s and A_s in quasi-hermitian. The vector u''_s belongs to the right ground of A_s and therefore every vector of the null space of A_s lies in the null space of A_{s+1}; also

$$A_{s+1} P_s u_s = A_s P_s u_s - u_s \frac{S u''_s P_s u_s}{S u''_s P_s u_s}$$

$$= u_s - u_s = 0.$$

Hence the null space of A_{s+1} is derived from that of A_s by adding $P_s u_s$ to it. It then follows as in §6.06 that A can be expressed in the form

(29) $$A = \sum_1^r \frac{u_s S u''_s}{S u''_s P_s u_s}$$

which is analogous to (16) and may be used in its place in proving Theorem 12.

We may also note that, if Q is the matrix defined by $Q'e_j = x_j$ ($j = 1, 2, \cdots, n$), then

$$A = Q' \sum_1^r q_{ij} e_i S e_j Q'' = Q'BQ''$$

where B is the quasi-hermitian matrix $\sum_1^r q_{ij} e_i S e_j$. It may be shown by an argument similar to that used for hermitian matrices that a basis for the x's may be so chosen that Q is quasi-orthogonal provided A is real.

6.07 Cogredient transformation. If $SxAy$ and $SxBy$ are two bilinear forms, the second is said to be derived from the first by a cogredient transformation if there exists a non-singular matrix P such that $SxAy \equiv SPxBPy$, that is,

$$A = P'BP. \tag{30}$$

When this relation holds between A and B, we shall say they are *cogredient*.

From (30) we derive immediately $A' = P'B'P$ and therefore, if

$$R = \frac{A + A'}{2} = R', \qquad S = \frac{A - A'}{2} = -S',$$

$$U = \frac{B + B'}{2} = U', \qquad V = \frac{B - B'}{2} = -V',$$

then

$$R + \lambda S = P'(U + \lambda V)P$$

so that $R + \lambda S$ and $U + \lambda V$ are strictly equivalent.

Suppose conversely that we are given that $R + \lambda S$ and $U + \lambda V$, which are quasi-hermitian, are strictly equivalent so that there exist constant non-singular matrices p, q such that

$$R + \lambda S = p(U + \lambda V)q$$

or

$$R = pUq, \qquad S = pVq; \tag{31}$$

then, remembering that R and U are symmetric, S and V skew, we have

$$R = q'Up', \qquad S = q'Vp'.$$

Equating these two values of R and S, respectively, we get

$$(q')^{-1}pU = Up'q^{-1}, \qquad (q')^{-1}pV = Vp'q^{-1}$$

or, if W stands for U or V indifferently, and

$$J = (q')^{-1}p, \tag{32}$$

we have

$$JW = WJ',$$

repeated application of which gives

$$J^r W = W(J')^r.$$

From this it follows that, if $f(\lambda)$ is a scalar polynomial,

$$f(J)W = Wf(J') = W(f(J))' \tag{33}$$

In particular, since $|J| \neq 0$, we may choose $f(\lambda)$ so that $f(J)$ is a square root of J and, denoting this square root by K, we have $KW = WK'$ or

$$W = K^{-1}WK', \qquad K^2 = J, \qquad (W = U \text{ or } V).$$

Using this in (31) we have

$$R = pK^{-1}UK'q, \qquad S = pK^{-1}VK'q$$

and from (32) $p = q'J = q'K^2$ or

$$pK^{-1} = q'K = (K'q)'.$$

Hence, if we put $P' = q'K$, there follows

$$R = P'UP, \qquad S = P'VP$$

or

$$A = R + S = P'(U + V)P = P'BP.$$

We therefore have the following theorem, which is due to Kronecker.

THEOREM 15. *A necessary and sufficient condition that A and B be cogredient is that $A + \lambda A'$ and $B + \lambda B'$ shall be strictly equivalent.*

If A and B are symmetric, these polynomials become $A(1 + \lambda)$ and $B(1 + \lambda)$ which are always strictly equivalent provided the ranks of A and B are the same. Hence quadratic forms of the same rank are always cogredient, as is also evident from Theorem 12 which shows in addition that P may be taken real if the signatures are the same.

The determination of P from (31) is unaltered if we suppose S symmetrical instead of skew, or R skew instead of symmetrical. Hence

THEOREM 16. *If R, S, U, V are all symmetric or all skew, and if $R + \lambda S$ and $U + \lambda V$ are strictly equivalent, we can find a constant non-singular matrix P such that*

$$R + \lambda S = P'(U + \lambda V)P,$$

that is, the corresponding pairs of forms are cogredient.

In the case of a hermitian form $S\bar{x}Ax$ changing x into Px replaces A by $\bar{P}'AP$ and we have in place of (30)

$$A = \bar{P}'BP. \tag{34}$$

If we put $B = \Sigma\beta_s b_s S\bar{b}_s$, then

$$\bar{P}'BP = \Sigma\beta_s\bar{P}'b_sS\bar{b}_sP = \Sigma\beta_s\bar{P}'b_sSP'\bar{b}_s = \Sigma\beta_sc_sS\bar{c}_s,$$

where $c_s = \bar{P}'b_s$. Equation (34) can therefore hold only if the signature as well as the rank is the same for B as for A. Conversely, if $A = \Sigma\alpha_sa_sS\bar{a}_s$ and A and B have the same signature and rank the notation may be so arranged that α_s and β_s have the same signs for all s; then any matrix for which

$$\bar{P}'b_s = \left(\frac{\alpha_s}{\beta_s}\right)^{\frac{1}{2}} a_s \quad (s = 1, 2, \cdots, r), \qquad |P| \neq 0$$

where r is the common rank of A and B, clearly satisfies (33).[2] Hence

THEOREM 17. *Two hermitian forms are cogredient if, and only if, they have the same rank and signature.*

The reader will readily prove the following extension of Theorem 16 by the aid of the artifice used in the proof of Theorem 11.

THEOREM 18. *If A, B, C, D are hermitian matrices such that $A + \lambda B$ and $C + \lambda D$ are* (i) *equivalent* (ii) *both definite for some value of λ, there exists a constant non-singular matrix P such that*

$$A + \lambda B = \bar{P}'(C + \lambda D)P.$$

6.08 **Real representation of a hermitian matrix.** Any matrix $H = A + iB$ in which A and B are real matrices of order n can be represented as a real matrix of order $2n$. For the matrix of order 2

$$i_2 = \left\| \begin{matrix} 0 & -1 \\ 1 & 0 \end{matrix} \right\|$$

satisfies the equation $i_2^2 = -1$ and, on forming the direct product of the original set of matrices of order n and a set of order 2 in which i_2 lies, we get a set of order $2n$ in which H is represented by

$$\mathfrak{H} = A + i_2B = \left\| \begin{matrix} A & -B \\ B & A \end{matrix} \right\|.$$

As a verification of this we may note that

$$\left\| \begin{matrix} A & -B \\ B & A \end{matrix} \right\| \left\| \begin{matrix} C & -D \\ D & C \end{matrix} \right\| = \left\| \begin{matrix} AC - BD & -(AD + BC) \\ AD + BC & AC - BD \end{matrix} \right\|$$

which corresponds to

$$(A + iB)(C + iD) = AC - BD + i(AD + BC).$$

[2] The proof preceding Theorem 15 generalizes readily up to equation (33); at that point, however, if $K = f(J)$, we require $\bar{K}' = f(\bar{J}')$, which is only true when the coefficients of $f(\lambda)$ are real.

This representation has the disadvantage that a complex scalar $\alpha + i\beta$ is represented by

$$\begin{Vmatrix} \alpha & -\beta \\ \beta & \alpha \end{Vmatrix}$$

which is not a scalar matrix although it is commutative with every matrix of the form $\mathfrak{H}$. Consequently, if H has a complex root, this root does not correspond to a root of $\mathfrak{H}$. If, however, all the roots of H are real, the relation $HK = \alpha K$ is represented by $\mathfrak{H}\mathfrak{K} = \alpha\mathfrak{K}$ when α is real so that α is a root of both H and $\mathfrak{H}$.

To prove the converse of this it is convenient to represent the vector $x + iy$ in the original space by (x, y) in the extended space. Corresponding to

$$(A + iB)(x + iy) = Ax - By + i(Bx + Ay)$$

we then have

$$\begin{Vmatrix} A & -B \\ B & A \end{Vmatrix} (x, y) = (Ax - By, Bx + Ay).$$

If therefore $\mathfrak{H}$ has a real root α and (x, y) is a corresponding invariant vector so that

$$\mathfrak{H}(x, y) = \alpha(x, y) = (\alpha x, \alpha y),$$

we have

$$Ax - By = \alpha x, \qquad Bx + Ay = \alpha y,$$

which gives

$$(A + iB)\,(x + iy) = \alpha(x + iy).$$

It follows that invariant vectors in the two representations correspond provided they belong to real roots. This gives

THEOREM 19. *To every real root of $H = A + iB$ there corresponds a real root of*

$$\mathfrak{H} = \begin{Vmatrix} A & -B \\ B & A \end{Vmatrix}$$

and vice-versa.

In this representation $\bar{H}$ and H' correspond to

$$\begin{Vmatrix} A & B \\ -B & A \end{Vmatrix}; \qquad \begin{Vmatrix} A' & -B' \\ B' & A' \end{Vmatrix},$$

respectively, and hence, if H is hermitian, $B' = -B$ so that $\mathfrak{H}$ is symmetric. The theory of hermitian matrices of order n can therefore be made to depend on that of real symmetric matrices of order $2n$. For example, if we have proved of real symmetric matrices that they have real roots and simple elementary divisors, it follows that the same is true of hermitian matrices, thus reversing the order of the argument made in §6.01.

CHAPTER VII

COMMUTATIVE MATRICES

7.01 We have already seen in §2.08 how to find all matrices commutative with a given matrix x which has no repeated roots. We shall now treat the somewhat more complicated case in which x is not so restricted. If

$$(1) \qquad xy = yx$$

then $x^r y = yx^r$ so that, if $f(\lambda)$ is a scalar polynomial, then $f(x)y = yf(x)$. In particular, if f_i is a principal idempotent element of x, then $f_i y = yf_i$. Remembering that $\Sigma f_i = 1$ we may set

$$x = \Sigma f_i x = \Sigma x_i, \qquad y = \Sigma f_i y = \Sigma y_i,$$

and also, by §2.11, $x_i = \lambda_i f_i + z_i$, where z_i is nilpotent. Since $y_i x_j = 0 = x_i y_j$ $(i \neq j)$, the determination of all matrices y which satisfy (1) is reduced to finding y so that

$$y_i x_i = x_i y_i, \qquad y_i = f_i y = yf_i.$$

We can therefore simplify the notation by first assuming that x has only one principal idempotent element, 1, and one root which may be taken to be 0 without loss of generality; x is then nilpotent.

Let $e_1, e_2, \cdots, e_s$ be the partial idempotent elements of x and let their ranks be $n_1, n_2, \cdots, n_s$; x is then composed of blocks of the form

$$\begin{matrix} 0 & 1 & 0 & \cdots & 0 \\ 0 & 0 & 1 & \cdots & 0 \\ 0 & 0 & 0 & \cdots & 0 \\ \cdots & \cdots & \cdots & \cdots & \cdots \\ 0 & 0 & 0 & \cdots & 1 \\ 0 & 0 & 0 & \cdots & 0 \end{matrix} \qquad (n_i \text{ rows and columns})$$

provided the fundamental basis is suitably chosen. To simplify the notation further we divide the array of n^2 units e_{ij} into smaller arrays formed by separating off the first n_1 rows, then the next n_2 rows, and so on, and then making a similar division of the columns (see figure 1). And when this is done, we shall denote the units in the block in which the ith set of rows meets the jth set of columns by

$$e_{pq}^{ij} \qquad (i, j = 1, 2, \cdots, s; p = 1, 2, \cdots, n_i; q = 1, 2, \cdots, n_j).$$

It is also convenient to put $e_{p\,q}^{i\,j} = 0$ for $p > n_i$ or $q > n_j$.

	n_1	n_2	n_3	$\cdots$
n_1	11	12	13	$\cdots$
n_2	21	22	23	$\cdots$
n_3	31	32	33	$\cdots$
	.	.	.	$\cdots$

FIG. 1

The expression for x is now

$$x = \sum_{i=1}^{s} \sum_{p=1}^{n_i - 1} e_{p,p+1}^{i\,i} = \sum_{i=1}^{s} x_i$$

and we may set

$$y = \sum_{i,j,p,q} \eta_{p\,q}^{i\,j} e_{p\,q}^{i\,j} = \sum_{i,j} y_{ij}$$

where

$$y_{ij} = e_i y e_j = \sum_{p=1}^{n_i} \sum_{q=1}^{n_j} \eta_{p\,q}^{i\,j} e_{p\,q}^{i\,j}.$$

The equation $xy = yx$ is then equivalent to

$$x_i y_{ij} = y_{ij} x_j \qquad (i, j = 1, 2, \cdots, s). \tag{2}$$

If we now suppress for the moment the superscripts i, j, which remain constant in a single equation in (2), we may replace (2) by

$$\sum_{p=1}^{n_i - 1} e_{p,\,p+1} \sum_{l=1}^{n_i} \sum_{m=1}^{n_j} \eta_{lm} e_{lm} = \sum_{l=1}^{n_i} \sum_{m=1}^{n_j} \eta_{lm} e_{lm} \sum_{q=1}^{n_j - 1} e_{q,q+1}$$

or

$$\sum_{m=1}^{n_j} \sum_{p=1}^{n_i - 1} \eta_{p+1,\,m} e_{pm} = \sum_{l=1}^{n_i} \sum_{q=1}^{n_j - 1} \eta_{lq} e_{l,q+1}. \tag{3}$$

Equating corresponding coefficients then gives

$$\eta_{p+1,\,q+1} = \eta_{pq}. \tag{4}$$

Since $q \geq 1$ on the right of (3), it follows that $\eta_{p+1,1} = 0$ $(p = 1, 2, \cdots, n_i - 1)$ and, since $p \leq n_i - 1$ on the left, $\eta_{n_i q} = 0$ $(q = 1, 2, \cdots, n_j - 1)$ and hence from (4)

$$\eta_{p+t,\,t} = 0 = \eta_{n_i - t,\,q - t} \tag{5}$$

where $p = 0, 1, \cdots, n_i - t$, $q = t + 1, s + 2, \cdots, n_j - 1$, $t = 0, 1, \cdots$.

From (4) we see that in y_{ij} all coordinates in an oblique line parallel to the main diagonal of the original array have the same value; from the first part of (5) those to the left of the oblique AB through the upper left hand corner are zero, as are also those to the left of the oblique CD through the lower right hand corner; the coordinates in the other obliques are arbitrary except that, as already stated, the coordinates in the same oblique are equal by (4). This state of affairs is made clearer by figure 2 where all coordinates are 0 except those in the shaded portion.

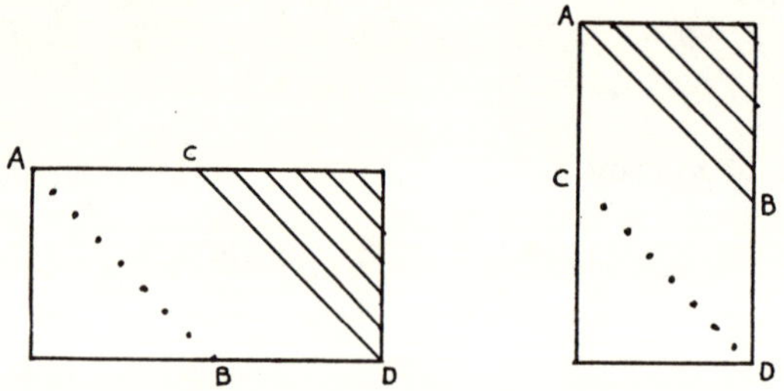

$n_i < n_j$ $\qquad$ $n_i > n_j$

FIG. 2

As an example of this take

$$x = \begin{array}{cccccccccc}
\alpha & 1 & & & & & & & & \\
 & \alpha & & & & & & & & \\
 & & \alpha & 1 & & & & & & \\
 & & & \alpha & 1 & & & & & \\
 & & & & \alpha & & & & & \\
 & & & & & \alpha & 1 & & & \\
 & & & & & & \alpha & 1 & & \\
 & & & & & & & \alpha & 1 & \\
 & & & & & & & & \alpha & 1 \\
 & & & & & & & & & \alpha
\end{array}$$

The above rules then give for y

$$(6) \qquad \begin{array}{cccccccccc}
a_0 & a_1 & . & b_0 & b_1 & . & . & . & c_0 & c_1 \\
. & a_0 & . & . & b_0 & . & . & . & . & c_0 \\
d_0 & d_1 & e_0 & e_1 & e_2 & . & . & f_0 & f_1 & f_2 \\
. & d_0 & . & e_0 & e_1 & . & . & . & f_0 & f_1 \\
. & . & . & . & e_0 & . & . & . & . & f_0 \\
g_0 & g_1 & h_0 & h_1 & h_2 & i_0 & i_1 & i_2 & i_3 & i_4 \\
. & g_0 & . & h_0 & h_1 & . & i_0 & i_1 & i_2 & i_3 \\
. & . & . & . & h_0 & . & . & i_0 & i_1 & i_2 \\
. & . & . & . & . & . & . & . & i_0 & i_1 \\
. & . & . & . & . & . & . & . & . & i_0
\end{array}$$

where the dots represent 0.

If we arrange the notation so that $n_1 \leq n_2 \leq \cdots \leq n_s$, a simple enumeration shows that the number of independent parameters in y is

$$(2s - 1)n_1 + (2s - 3)n_2 + \cdots + n_s.$$

We have therefore the following theorem which is due to Frobenius.

THEOREM 1. *If the elementary divisors of x are $(\lambda - \lambda_i)^{n_{ij}}$, $i = 1, 2, \cdots, r$, $j = 1, 2, \cdots, s_i$, where $\lambda_1, \lambda_2, \cdots, \lambda_r$ are all different and $n_{i1} \leq n_{i2} \leq \cdots \leq n_{is_i}$, then the general form of a matrix commutative with x depends on*

$$\sum_{i=1}^{r} \sum_{j=1}^{s_i} (2s - 2j + 1)n_{ij}$$

independent parameters.

7.02 Commutative sets of matrices. The simple condition $xy = yx$ may be replaced by the more stringent one that y is commutative with every matrix which is commutative with x. To begin with we shall merely assume that y is commutative with each of a particular set of partial idempotent elements e_i; as in the previous section we may assume that x has only one principal idempotent element.

In order that $e_i y = y e_i$ for every i it is necessary and sufficient that $y_{ij} = 0$ when $i \neq j$; if $u_1, u_2, \cdots, u_s$ are the partial nilpotent elements of x corresponding to $e_1, e_2, \cdots, e_s$ and we set $m_i = n_i - 1$, this gives for y

$$y = \sum_i (\eta_{i0} e_i + \eta_{i1} u_i + \cdots + \eta_{im_i} u_i^{m_i}). \tag{7}$$

If we now put $z = \sum_i (\beta_i e_i + u_i)$, where no $\beta_i = 0$, and if $g(\lambda)$ is any scalar polynomial, then (cf. §2.11)

$$g(z) = \Sigma g(\beta_i e_i + u_i) = \Sigma(g(\beta_i)e_i + g'(\beta_i)u_i + \cdots + g^{(m_i)}(\beta_i)u_i^{m_i}/m_i!)$$

and when y is given, we can always find $g(\lambda)$ so that

$$\eta_{ik} = g^{(k)}(\beta_i)/k!$$

provided the β's are all different. Hence every y, including x itself, can be expressed as a polynomial in z.

We now impose the more exacting condition that y is permutable with every matrix permutable with x. Let u_{ij} $(i \neq j)$ be the matrix of the same form as y_{ij} in §7.01 but with zero coordinates everywhere except in the principal oblique; for example in (6) u_{23} is obtained by putting $f_0 = 1$ and making every other coordinate 0. We then have

$$e_i u_{ij} = u_{ij} e_j, \qquad u_i u_{ij} = u_{ij} u_j.$$

Hence $yu_{ij} = u_{ij}y$ gives $y_{ii}u_{ij} = u_{ij}y_{jj}$ and therefore from (7)

$$\begin{aligned}(\eta_{i0}e_i + \eta_{i1}u_i + \cdots + \eta_{im_i}u_i^{m_i})u_{ij} &= u_{ij}(\eta_{j0}e_j + \eta_{j1}u_j + \cdots + \eta_{jm_j}u_j^{m_j}) \\ &= (\eta_{j0}e_i + \eta_{j1}u_i + \cdots + \eta_{jm_j}u_i^{m_j})u_{ij}\end{aligned}$$

from which we readily derive for all i, j and k

$$\eta_{ik} = \eta_{jk}$$

with the understanding that η_{ik} does not actually occur when $k > m_i$. When x is the matrix used in deriving (6), these conditions give in place of (6)

$$(8) \qquad \begin{array}{cccccccccc}
a_0 & a_1 & \cdot & \cdot & \cdot & \cdot & \cdot & \cdot & \cdot & \cdot \\
\cdot & a_0 & \cdot & \cdot & \cdot & \cdot & \cdot & \cdot & \cdot & \cdot \\
\cdot & \cdot & a_0 & a_1 & a_2 & \cdot & \cdot & \cdot & \cdot & \cdot \\
\cdot & \cdot & \cdot & a_0 & a_1 & \cdot & \cdot & \cdot & \cdot & \cdot \\
\cdot & \cdot & \cdot & \cdot & a_0 & \cdot & \cdot & \cdot & \cdot & \cdot \\
\cdot & \cdot & \cdot & \cdot & \cdot & a_0 & a_1 & a_2 & a_3 & a_4 \\
\cdot & \cdot & \cdot & \cdot & \cdot & \cdot & a_0 & a_1 & a_2 & a_3 \\
\cdot & \cdot & \cdot & \cdot & \cdot & \cdot & \cdot & a_0 & a_1 & a_2 \\
\cdot & \cdot & \cdot & \cdot & \cdot & \cdot & \cdot & \cdot & a_0 & a_1 \\
\cdot & \cdot & \cdot & \cdot & \cdot & \cdot & \cdot & \cdot & \cdot & a_0.
\end{array}$$

Comparing this form with (7) we see that y is now a scalar polynomial in x, which in the particular case given above becomes $g(x - \alpha)$ where

$$g(\lambda) = a_0 + a_1\lambda + a_2\lambda^2 + a_3\lambda^3 + a_4\lambda^4.$$

The results of this section may be summarized as follows.

Theorem 2. *Any matrix which is commutative, not only with x, but also with every matrix commutative with x, is a scalar polynomial in x.*

7.03 Rational methods. Since the solution of $xy - yx = 0$ for y can be regarded as equivalent to solving a system of linear homogeneous equations, the solution should be expressible rationally in terms of suitably chosen parameters; the method of §7.01, though elementary and direct, cannot therefore be regarded as wholly satisfactory. The following discussion, which is due to Frobenius, avoids this difficulty but is correspondingly less explicit.

As before let $xy = yx$ and set $a = \lambda - x$; also let $b = L^{-1}aM^{-1}$ be the normal form of a. If u is an arbitrary polynomial in λ and we set

$$P = L^{-1}(au + y)L, \qquad Q = M(ua + y)M^{-1},$$

then

$$Pb = PL^{-1}aM^{-1} = L^{-1}(au + y)aM^{-1} = L^{-1}a(ua + y)M^{-1} = bQ.$$

Conversely, if $Pb = bQ$ and, using the division transformation, we set

$$LPL^{-1} = av + y, \qquad M^{-1}QM = v_1a + y_1, \tag{9}$$

where y and y_1 are constants, then

$$0 = Pb - bQ = L^{-1}(av + y)aM^{-1} - L^{-1}a(v_1a + y_1)M^{-1}$$

or $a(v - v_1)a = ay_1 - ya$. Here the degree on the left is at least 2 and on the right only 1 and hence by the usual argument both sides of the equation vanish. This gives

$$ava = av_1a, \qquad ay_1 = ya$$

whence $v_1 = v$ and, since $a = \lambda - x$, also $y_1 = y$ so that $xy = yx$.

Hence we can find all matrices commutative with x by finding all solutions of

$$Pb = bQ. \tag{10}$$

Let $\alpha_1, \alpha_2, \cdots, \alpha_n$ be the invariant factors[1] of a and $n_1, n_2, \cdots, n_n$ the corresponding degrees so that b is the diagonal matrix $\Sigma\alpha_ie_{ii}$, and let $P = \| P_{ij} \|$, $Q = \| Q_{ij} \|$; then

$$P_{ij}\alpha_j = \alpha_iQ_{ij}. \tag{11}$$

By the division transformation we may set

$$P_{ij} = R_{ij}\alpha_i + p_{ij}, \qquad Q_{ij} = S_{ij}\alpha_j + q_{ij}$$

and then from (10) we have

$$R_{ij} = S_{ij}, \qquad p_{ij}\alpha_j = \alpha_iq_{ij}$$

or, if $p = \| p_{ij} \|$, $q = \| q_{ij} \|$,

$$pb = bq. \tag{12}$$

Hence $P = p$, $Q = q$ is a solution of (10) for which the degree of p_{ij} is less than that of α_i and the degree of q_{ij} is less than that of α_j. It is then evident that, when the general solution p, q of (12) is found, then the general solution of (10) has the form

$$P = bR + p, \qquad Q = Rb + q$$

where R is an arbitrary matric polynomial in λ. We are however not concerned with R; for

$$LPL^{-1} = LbRL^{-1} + LpL^{-1} = aM^{-1}RL^{-1} + LpL^{-1}$$

so that in (9) the value of y depends on p only.

[1] Since we may add a scalar to x we may clearly assume that the rank of a is n.

The general solution of (12) is given by

$$(13)\qquad \begin{aligned} p_{ij} &= \frac{\alpha_i}{\alpha_j} s_{ij}, & p_{ji} &= s_{ji} \\ q_{ij} &= s_{ij}, & q_{ji} &= \frac{\alpha_i}{\alpha_j} s_{ji} \end{aligned} \qquad (i > j)$$

where $s_{\lambda\mu}$ is an arbitrary polynomial whose degree is at most $n_\mu - 1$ and which therefore depends on n_μ parameters. It follows that the total number of parameters in the value of y is that already given in §7.01.

7.04 The direct product. We shall consider in this section some properties of the direct product which was defined in §5.10.

THEOREM 3. *If f_{ij} $(i, j = 1, 2, \cdots, m)$ is a set of matrices, of order n, for which*

$$(14)\qquad f_{ij}f_{pq} = \delta_{jp}f_{iq}, \qquad \sum_{i=1}^{m} f_{ii} = 1,$$

then m is a factor of n and any matrix of order n can be expressed uniquely in the form $\Sigma a_{ij}f_{ij}$ where each a_{ij} is commutative with every f_{pq}; and, if $n = mr$, the rank of each f_{pq} is r.

For, if x is an arbitrary matrix and we set

$$(15)\qquad a_{ij} = \sum_{k=1}^{m} f_{ki}xf_{jk},$$

a short calculation shows:

(i) $x = \Sigma a_{ij}f_{ij}$;

(ii) $a_{ij}f_{pq} = f_{pq}a_{ij}$ for all i, j, p, q;

(iii) the set $\mathfrak{A}$ of all matrices of the form (15) is closed under the operations of addition and multiplication;

(iv) if $b_{11}, b_{12}, \cdots$ are members of $\mathfrak{A}$, then $\Sigma b_{ij}f_{ij}$ is zero if, and only if, each $b_{ij} = 0$.

If $(a_1, a_2, \cdots, a_l)$ is a basis of $\mathfrak{A}$, it follows that

$$(a_pf_{ij}\colon p = 1, 2, \cdots, l; i, j = 1, 2, \cdots, m)$$

is equivalent to the basis $(e_{ij}, i, j = 1, 2, \cdots, n)$ of the set of matrices of order n. This basis contains lm^2 independent elements and hence $n^2 = lm^2$ so that $n = mr$, $l = r^2$. Let r_{ij} be the rank of f_{ij}. Since $f_{ii} = f_{ij}f_{ji}$, it follows from Theorem 8 of chapter I that $r_{ii} \leq r_{ji}$; also from $f_{ji}f_{ii} = f_{ji}$ we have $r_{ji} \leq r_{ii}$; hence $r_{ii} = r_{ji}$ and therefore each r_{ji} has the same value. Finally, since $1 = \Sigma f_{ii}$, and $f_{ii}f_{jj} = 0$ $(i \neq j)$ and $r_{ii} = r_{jj}$, we have $mr_{ii} = n$ and hence each $r_{ij} = r$.

We shall now show that a basis g_{ij} can be chosen for $\mathfrak{A}$ which satisfies the relations (14) with r in place of m. Since the rank of f_{ii} is r, we can set

$$f_{ii} = \sum_1^r \alpha_{ik} S \beta_{ik} \qquad (i = 1, 2, \cdots, m) \tag{16}$$

where the sets of vectors (α_{ik}) and (β_{ik}) $(i = 1, 2, \cdots, m;\ k = 1, 2, \cdots, r)$ each form a basis of the n-space since $\sum_{i=1}^m f_{ii} = 1$. If (α'_{ik}), (β'_{ik}) are the corresponding reciprocal sets and

$$p_{ii} = \sum_1^r \beta'_{ik} S \alpha'_{ik} \qquad (i = 1, 2, \cdots, m)$$

we have, since $S\alpha_{ik}\alpha'_{js} = \delta_{ij}\delta_{ks}$,

$$\Sigma p_{ii} = \Sigma f_{ii} \Sigma p_{ii} = \Sigma\,\Sigma \alpha_{ik} S \beta_{ik} \beta'_{ij} S \alpha'_{ij} = \Sigma \alpha_{ik} S \alpha'_{ik} = 1,$$

and similarly

$$f_{ii} p_{ii} = \sum_k \alpha_{ik} S \alpha'_{ik}, \qquad f_{ii} p_{jj} = 0 \qquad (i \neq j).$$

Hence

$$f_{ii} = f_{ii} \Sigma p_{jj} = \sum_k \alpha_{ik} S \alpha'_{ik}, \tag{17}$$

that is $\beta_{ik} = \alpha'_{ik}$.

Since $f_{ij} = f_{ii} f_{ij} f_{jj}$, the left ground of f_{ij} is the same as that of f_{ii} and its right ground is the same as that of f_{jj}. Let

$$f_{1j} = \Sigma \alpha_{1k} S \gamma_{jk}.$$

The vectors γ_{jk} $(k = 1, 2, \cdots, r)$ then form a basis for the set α'_{jk} $(k = 1, 2, \cdots, r)$ and, since the basis chosen for this set in (16) is immaterial, we may suppose $\gamma_{jk} = \alpha'_{jk}$ $(j = 1, 2, \cdots, m;\ k = 1, 2, \cdots, r)$, that is,

$$f_{1j} = \sum_k \alpha_{1k} S \alpha'_{jk}.$$

Similarly we may set $f_{j1} = \Sigma \alpha_{jk} S \theta_{1k}$ and then since

$$\sum_k \alpha_{1k} S \alpha'_{1k} = f_{11} = f_{1j} f_{j1} = \sum_{k,s} \alpha_{1k} S \alpha'_{jk} \alpha_{js} S \theta_{1s} = \Sigma \alpha_{1k} S \theta_{1k},$$

we have $\theta_{1k} = \alpha'_{1k}$ and therefore

$$f_{j1} = \Sigma \alpha_{jk} S \alpha'_{1k}$$

and finally

$$f_{ij} = f_{i1}f_{1j} = \sum_{k,s} \alpha_{ik}S\alpha'_{1k}\alpha_{1s}S\alpha'_{js} = \Sigma\alpha_{ik}S\alpha'_{jk}. \tag{18}$$

If we now set $\alpha_{ik} = Pe_{(i-1)r+k}$, then by §1.09 P is non-singular and $\alpha'_{ik} = (P')^{-1}e_{(i-1)r+k}$; hence, if

$$h_{ij} = \sum_{k=1}^{r} e_{(i-1)r+k,\ (j-1)r+k}, \tag{19}$$

we have

$$f_{ij} = Ph_{ij}P^{-1}. \tag{20}$$

Also if

$$k_{ij} = \sum_{s=0}^{m-1} e_{sr+i,\ sr+j}, \tag{21}$$

then

$$k_{ij}h_{p+1,\ q+1} = e_{pr+i,\ qr+j} = h_{p+1,\ q+1}k_{ij} \tag{22}$$

so that the set (e_{ij}) of all matrices of order n may be regarded as the direct product of the sets (h_{ij}) and (k_{ij}). Finally, since any matrix can be expressed in the form $\Sigma b_{ij}h_{ij}$, where the b_{ij} depend on the basis (k_{ij}), it follows that an arbitrary matrix can also be expressed in the form

$$P\Sigma b_{ij}h_{ij}P^{-1} = \Sigma Pb_{ij}P^{-1}f_{ij};$$

$Pb_{ij}P^{-1}$ depends on the basis $(Pk_{ij}P^{-1})$ and hence, if we set

$$g_{ij} = Pk_{ij}P^{-1} \qquad (i, j = 1, 2, \cdots, r)$$

the g's form a basis of $\mathfrak{A}$ which satisfies (14).

7.05 Functions of commutative matrices. Let x and y be commutative matrices whose distinct roots are $\lambda_1, \lambda_2, \cdots$ and $\mu_1, \mu_2, \cdots$ respectively and let R_i be the principal idempotent unit of x corresponding to λ_i and similarly S_j the principal idempotent unit of y corresponding to μ_j. Since R_i and S_j are scalar polynomials in x and y, they are commutative. If we set

$$T_{ij} = R_iS_j,$$

those T_{ij} which are not 0 are linearly independent; for if $\Sigma\xi_{ij}T_{ij} = 0$, then

$$0 = R_p\Sigma\xi_{ij}T_{ij}S_q = \xi_{pq}T_{pq},$$

since $R_pR_i = \delta_{pi}R_p$, $S_jS_q = \delta_{jq}S_q$, so that either $\xi_{pq} = 0$ or $T_{pq} = 0$.

From the definition of T_{ij} it follows that $T_{ij}T_{pq} = 0$ when $i \neq p$ or $j \neq q$, and $T_{ij}^2 = T_{ij}$, $\Sigma T_{ij} = 1$; hence

$$x = \sum_{i,j} [\lambda_i + (x - \lambda_i)]T_{ij}, \qquad y = \sum_{i,j} [\mu_j + (y - \mu_j)]T_{ij},$$

where $(x - \lambda_i)T_{ij}$ and $(y - \mu_j)T_{ij}$ are nilpotent. If $\psi(\lambda, \mu)$ is any scalar polynomial then

$$\psi(\lambda\mu) = \psi(\lambda_i, \mu_j) + \sum_{r,s} \psi_{rs}^{ij}(\lambda - \lambda_i)^r(\mu - \mu_j)^s$$

where ψ_{rs}^{ij} are scalars, we have therefore

$$\begin{aligned}\psi(x, y) &= \sum_{i,j} \left[\psi(\lambda_i, \mu_j)T_{ij} + \sum_{r,s} \psi_{rs}^{ij}(x - \lambda_i)^r(y - \mu_j)^s T_{ij}\right] \\ &= \sum_{i,j} \psi(\lambda_i, \mu_j)T_{ij} + \sum_{i,j}\sum_{r,s} \psi_{rs}^{ij}T_{ij}^{rs}\end{aligned}$$

where

$$T_{ij}^{rs} = (x - \lambda_i)^r(y - \mu_j)^s T_{ij}$$

and r runs from 1 to $\rho_i - 1$, where ρ_i is the smallest integer for which $(x - \lambda_i)^{\rho_i}R_i = 0$, and s has a similar range with respect to y. The matrices T_{ij}^{rs} are commutative and each is nilpotent; and hence any linear combination of them is also nilpotent.

Let

$$z = \Sigma\psi(\lambda_i, \mu_j)T_{ij}, \qquad w = \Sigma\Sigma\psi_{rs}^{ij}T_{ij}^{rs};$$

then w, being the sum of commutative nilpotent matrices, is nilpotent. If we take in z only terms for which $T_{ij} \neq 0$, we see immediately that the roots of z are the corresponding coefficients $\psi(\lambda_i, \mu_i)$; and the reduced characteristic function of z is found as in §2.12. We have therefore the following theorem which is due to Frobenius.

Theorem 4. *If R_i, S_j $(i = 1, 2, \cdots; j = 1, 2, \cdots)$ are the principal idempotent units of the commutative matrices x, y and $T_{ij} = R_iS_j$; and if λ_i, μ_j are the corresponding roots of x and y, respectively; then the roots of any scalar function $\psi(x, y)$ of x and y are $\psi(\lambda_i, \mu_j)$ where i and j take only those values for which $T_{ij} \neq 0$.*

This theorem extends immediately to any number of commutative matrices.

7.06 Sylvester's identities. It was shown in §2.08 that, if the roots of x are all distinct, the only matrices commutative with it are scalar polynomials in x; and in doing so certain identities, due to Sylvester, were derived. We shall now consider these identities in more detail.

We have already seen that in

$$f(\lambda) = |\lambda - x| = \lambda^n + a_1\lambda^{n-1} + \cdots + a_{n-1}\lambda + a_n$$

the coefficient a_r of λ^{n-r} is $(-1)^r$ times the sum of the principal minors of x of order r; these coefficients are therefore homogeneous polynomials of degree r in the coordinates of x. We shall now denote $(-1)^r a_r$ by $\left[\begin{smallmatrix} x \\ r \end{smallmatrix}\right]$. If x is replaced by $\lambda x + \mu y$, then $\left[\begin{smallmatrix} x \\ r \end{smallmatrix}\right]$ can be expressed as a homogeneous polynomial in λ, μ of degree r, and we shall write

$$(23) \qquad \begin{bmatrix} \lambda x + \mu y \\ r \end{bmatrix} = \sum_{s=0}^{r} \begin{Bmatrix} x & y \\ s & r-s \end{Bmatrix} \lambda^s \mu^{r-s}.$$

We shall further set, as in §2.08,

$$(24) \qquad (\lambda x + \mu y)^r = \sum_{s=0}^{r} \begin{Bmatrix} x & y \\ s & r-s \end{Bmatrix} \lambda^s \mu^{r-s}$$

where $\left\{\begin{smallmatrix} x & y \\ s & t \end{smallmatrix}\right\}$ is obtained by multiplying s x's and t y's together in every possible way and adding the terms so obtained.

In this notation the characteristic equation of $\lambda x + \mu y$ is

$$(25) \qquad 0 = \sum_{r=0}^{n} (-1)^r \begin{bmatrix} \lambda x + \mu y \\ r \end{bmatrix} (\lambda x + \mu y)^{n-r}$$

$$= \sum_{r,s,t} \begin{bmatrix} x & y \\ s & r-s \end{bmatrix} \begin{Bmatrix} x & y \\ t-s & n+s-r-t \end{Bmatrix} \lambda^t \mu^{n-t},$$

where in the second summation $\left[\begin{smallmatrix} x & y \\ p & q \end{smallmatrix}\right]$ or $\left\{\begin{smallmatrix} x & y \\ p & q \end{smallmatrix}\right\}$ is to be replaced by 0 if either p or q is negative and $\left[\begin{smallmatrix} x & y \\ 0 & 0 \end{smallmatrix}\right] = 1$. Since λ is an independent variable, the coefficients of its various powers in (25) are identically 0, and therefore

$$(26) \qquad \sum_{r,s=0}^{n} (-1)^r \begin{bmatrix} x & y \\ s & r-s \end{bmatrix} \begin{Bmatrix} x & y \\ t-s & n+s-r-t \end{Bmatrix} = 0 \qquad (t = 0, 1, \cdots, n)$$

a series of identical relations connecting two arbitrary matrices.

These identities can be generalized immediately. If $x_1, x_2, \cdots, x_m$ are any matrices and $\lambda_1, \lambda_2, \cdots,$ scalar variables, we may write

$$(27) \qquad \begin{aligned} \begin{bmatrix} \Sigma\lambda_i x_i \\ r \end{bmatrix} &= \sum \begin{bmatrix} x_1 & x_2 & \cdots & x_m \\ r_1 & r_2 & \cdots & r_m \end{bmatrix} \lambda_1^{r_1} \lambda_2^{r_2} \cdots \lambda_m^{r_m} \\ (\Sigma\lambda_i x_i)^r &= \sum \begin{Bmatrix} x_1 & x_2 & \cdots & x_m \\ r_1 & r_2 & \cdots & r_m \end{Bmatrix} \lambda_1^{r_1} \lambda_2^{r_2} \cdots \lambda_m^{r_m} \end{aligned} \qquad (\Sigma r_i = r)$$

and by the same reasoning as before we have

$$(28) \qquad \sum_r \sum_{r_1 \cdots r_m} (-1)^r \begin{bmatrix} x_1 & x_2 & \cdots & x_m \\ r_1 & r_2 & \cdots & r_m \end{bmatrix} \begin{Bmatrix} x_1 & x_2 & \cdots & x_m \\ s_1 - r_1 & s_2 - r_2 & \cdots & s_m - r_m \end{Bmatrix} = 0$$

$$\left(\sum_1^m r_i = r \right)$$

where $s_1, s_2, \cdots, s_m$ is any partition of n, zero parts included, and as before a bracket symbol is 0 when any exponent is negative.

Since $\begin{bmatrix} \Sigma\lambda_i x_i \\ r \end{bmatrix}$ is the sum of the principal minors of $\Sigma\lambda_i x_i$ of order r, we see that $\begin{bmatrix} x_1 & x_2 & \cdots & x_m \\ r_1 & r_2 & \cdots & r_m \end{bmatrix}$ $(\Sigma r_i = r)$ is formed as follows. Take any principal minor of x_1 of order r and the corresponding minors of $x_2, x_3, \cdots, x_m$ and replace r_2 of its columns by the corresponding columns of x_2, then replace r_3 of the remaining columns by the corresponding ones of x_3, and so on; do this in every possible way for each of the minors of order r of x_1 and add all the terms so obtained.

There is a great variety of relations connecting the scalar functions defined above, a few of which we note here for convenience.

(i)
$$\begin{bmatrix} 1 \\ r \end{bmatrix} = \frac{n!}{r!(n-r)!}, \qquad \begin{bmatrix} x & 1 \\ r & s \end{bmatrix} = \frac{(n-r)!}{s!(n-r-s)!}\begin{bmatrix} x \\ r \end{bmatrix},$$
$$\begin{bmatrix} x \\ n \end{bmatrix} = |\, x \,|, \qquad \begin{bmatrix} x & x \\ r & s \end{bmatrix} = \frac{(r+s)!}{r!s!}\begin{bmatrix} x \\ r+s \end{bmatrix}.$$

(ii) The value of $\begin{bmatrix} x_1 & x_2 & \cdots & x_m \\ & & r & \end{bmatrix}$ is unchanged by a cyclic permutation of the x's.

(iii)
$$\begin{bmatrix} 1 & 1 & \cdots & 1 \\ r_1 & r_2 & \cdots & r_m \end{bmatrix} = \frac{n!}{\Pi(r_i!)\,(n-\Sigma r_i)!},$$
$$\begin{bmatrix} x_1 & x_2 & \cdots & x_m & 1 \\ r_1 & r_2 & \cdots & r_m & s \end{bmatrix} = \frac{(n-\Sigma r_i)!}{s!(n-s-\Sigma r_i)!}\begin{bmatrix} x_1 & x_2 & \cdots & x_m \\ r_1 & r_2 & \cdots & r_m \end{bmatrix}$$
$$\begin{bmatrix} x & x & \cdots & x & y_1 & \cdots & y_p \\ r_1 & r_2 & \cdots & r_m & s_1 & \cdots & s_p \end{bmatrix} = \frac{(\Sigma r_i)!}{\Pi(r_i!)}\begin{bmatrix} x & y_1 & \cdots & y_p \\ \Sigma r_i & s_1 & \cdots & s_p \end{bmatrix}.$$

(iv)
$$\begin{bmatrix} x \\ r \end{bmatrix}\begin{bmatrix} y_1 & y_2 & \cdots & y_n \\ 1 & 1 & \cdots & 1 \end{bmatrix} = \Sigma\begin{bmatrix} xy_{i_1} & xy_{i_2} & \cdots & xy_{i_r} & y_{i_{r+1}} & \cdots & y_{i_n} \\ 1 & 1 & \cdots & 1 & 1 & \cdots & 1 \end{bmatrix}$$

where the summation extends over the $n!/r!(n-r)!$ ways of choosing r integers out of $1, 2, \cdots, n$, the order being immaterial.

7.07 **Similar matrices.** In addition to the identities discussed in the preceding section Sylvester gave another type, a modification of which we shall now discuss. If x, y, a are arbitrary matrices, we have

$$(29) \qquad x^{r+1}a - ay^{r+1} = x(x^r a + x^{r-1}ay + x^{r-2}ay^2 + \cdots + ay^r) - (x^r a + x^{r-1}ay + x^{r-2}ay^2 + \cdots + ay^r)y$$

or say

$$x^{r+1}a - ay^{r+1} = x(x, a, y)_r - (x, a, y)_r y$$

where

(30) $$(x, a, y)_r = \sum_{i=0}^{r} x^{r-i} a y^i.$$

Suppose now that x and y satisfy the same equation $f(\lambda) = 0$ where

$$f(\lambda) = \lambda^m + a_1\lambda^{m-1} + \cdots + a_{m-1}\lambda + a_m$$

x and y being commutative with each a_i and a commutative with every a_i. Let

(31) $$u = \sum_{0}^{m-1} a_i(x, a, y)_{m-i-1};$$

then

(32) $$0 = f(x)a - af(y) = \Sigma a_i(x^{m-i} a - ay^{m-i}) = xu - uy.$$

If $| u | \neq 0$, it follows that $y = u^{-1}xu$, that is, x and y are similar.

It can be shown that a can be chosen so that $| u | \neq 0$ provided x and y have the same invariant factors and $f(\lambda)$ is the reduced characteristic function.

CHAPTER VIII

FUNCTIONS OF MATRICES

8.01 Matric polynomials. The form of a polynomial in a matrix has already been discussed in §2.11 but we repeat the principal formulas here for convenience. If x is a matrix whose reduced characteristic function is

$$\varphi(\lambda) = \prod_{i=1}^{r} (\lambda - \lambda_i)^{\nu_i}, \qquad \Sigma\nu_i = \nu, \tag{1}$$

and

$$m_i(\lambda) = \varphi(\lambda)/(\lambda - \lambda_i)^{\nu_i}, \qquad M_i(\lambda)m_i(\lambda) + (\lambda - \lambda_i)^{\nu_i}N_i(\lambda) \equiv 1,$$

$$\varphi_i(\lambda) = M_i(\lambda)m_i(\lambda), \tag{2}$$

$$\varphi_i = \varphi_i(x), \qquad h_i = (x - \lambda_i)\varphi_i, \tag{3}$$

and if $g(\lambda)$ is a scalar polynomial in λ, then

$$g(x) = \sum_{i=1}^{r} \left[g(\lambda_i)\varphi_i + g'(\lambda_i)h_i + \cdots + \frac{g^{(\nu_i - 1)}(\lambda_i)h_i^{\nu_i - 1}}{(\nu_i - 1)!} \right]. \tag{4}$$

This formula can still be interpreted when the coefficients of $g(\lambda)$ are matrices, but in this case the notation $g(x)$ is ambiguous. Let $g(\lambda) = a_0 + a_1\lambda + \cdots + a_m\lambda^m$; then

$$a_0 + a_1x + \cdots + a_mx^m \quad \text{and} \quad a_0 + xa_1 + \cdots + x^ma_m$$

are called, respectively, the *dextro-* and *laevo-lateral* polynomials corresponding to $g(\lambda)$. It is clear that (4) holds for a dextro-lateral polynomial and will give the corresponding laevo-lateral polynomial if $g(\lambda_i)\varphi_i$, $g'(\lambda_i)h_i$, etc., are replaced by $\varphi_i g(\lambda_i)$, $h_i g'(\lambda_i)$, etc.

8.02 Infinite series. If $a_0, a_1, \cdots$ are matrices and λ a scalar variable, the coordinates of the matrix

$$g(\lambda) = a_0 + a_1\lambda + a_2\lambda^2 + \cdots \tag{5}$$

are scalar infinite series in λ; and if each of these series converges for mod λ less than ρ, we say that the series (5) converges. When this condition is satisfied, the series

$$g(x) = a_0 + a_1x + a_2x^2 + \cdots \tag{6}$$

converges for any matrix x for which the absolute value of the root of greatest absolute value is less than ρ. For if g_m is the sum of the first m terms of (6), then by (4) $g_m = \sum_i g_{mi}$ where

$$g_{mi} = g_m(\lambda_i)\varphi_i + g_m'(\lambda_i)h_i + \cdots + \frac{g_m^{(\nu_i - 1)}(\lambda_i)h_i^{\nu_i - 1}}{(\nu_i - 1)!}.$$

The matrices φ_i, h_i are independent of m and, since the absolute value of each λ_i is less than ρ, $g_m(\lambda_i)$, $g_m'(\lambda_i)$, $\cdots$, $g_m^{(\nu_i - 1)}(\lambda_i)$ converge to $g(\lambda_i)$, $g'(\lambda_i)$, $\cdots$, $g^{(\nu_i - 1)}(\lambda_i)$ when m approaches infinity.

As an illustration of such a series we may define $\exp x$ and $\log (1 + x)$ by

$$\text{(7)} \qquad \exp x = \epsilon^x = 1 + \frac{x}{1} + \frac{x^2}{2!} + \cdots$$

$$\text{(8)} \qquad \log (1 + x) = x - \frac{x^2}{2} + \frac{x^3}{3} + \cdots$$

The first of these converges for every matrix x, the second for matrices all of whose roots are less than 1 in absolute value.

The usual rules for adding series and for multiplying series whose coefficients are commutative with x and with each other hold for matric series. For instance we can show by the ordinary proof that, if $xy = yx$, then $\epsilon^{x+y} = \epsilon^x\epsilon^y$ but this will not usually be the case if $xy \neq yx$.

8.03 The canonical form of a function. In the case of multiform functions (4) does not always give the most general determination of the function which is only obtained by taking into account the partial as well as the principal elements of the variable x. As in §3.06 suppose that x has the canonical form

$$\text{(9)} \qquad x = \left\| \begin{matrix} a_1 & & & & \\ & a_2 & & & \\ & & \cdot & & \\ & & & \cdot & \\ & & & & \cdot \\ & & & & a_r \end{matrix} \right\|$$

where a_i is a block of terms

$$\text{(10)} \qquad \begin{matrix} \lambda_i & 1 & 0 & \cdots & 0 & 0 \\ 0 & \lambda_i & 1 & \cdots & 0 & 0 \\ \cdots & \cdots & \cdots & \cdots & \cdots & \cdots \\ \cdots & \cdots & \cdots & \cdots & \cdots & \cdots \\ 0 & 0 & 0 & \cdots & \lambda_i & 1 \\ 0 & 0 & 0 & \cdots & 0 & \lambda_i \end{matrix} \qquad (r_i \text{ rows and columns}).$$

It is convenient to let a_i stand also for the matrix derived from (9) by replacing every a_j by 0 except when $j = i$. We can then write

$$x = \Sigma a_i, \qquad a_i a_j = 0 \qquad (i \neq j) \tag{11}$$

and we may set

$$a_i = \lambda_i e_i + z_i \tag{12}$$

where (cf. §3.07) $e_i^2 = e_i$, z_i is a nilpotent matrix of index r_i, and

$$\sum_i e_i = 1, \qquad e_i z_i = z_i = z_i e_i, \qquad e_i e_j = 0, \qquad e_i z_j = 0 = z_j e_i \qquad (i \neq j).$$

The part of z_i which is not 0 is given by the oblique line of 1's in (10); z^2 is obtained by moving all the 1's one place to the right except the last which disappears, and in general z^{m-1} has a line of 1's starting in the mth column of (10) and running parallel to the main diagonal till it meets the boundary of the block.

It is now easy to see the form of a scalar polynomial $g(x)$ or of a convergent power series with scalar coefficients; for

$$g(x) = \Sigma g(a_i) e_i = \Sigma \left[g(\lambda_i) e_i + g'(\lambda_i) z_i + \cdots + \frac{g^{(r_i - 1)}(\lambda_i) z_i^{r_i - 1}}{(r_i - 1)!} \right] \tag{13}$$

and the block of terms in $g(x)$ which corresponds to a_i in (10) is, omitting the subscripts for clearness,

$$\begin{array}{cccccc} g(\lambda) & g'(\lambda) & \dfrac{g''(\lambda)}{2!} & \dfrac{g'''(\lambda)}{3!} & \cdots & \dfrac{g^{(r-1)}(\lambda)}{(r-1)!} \\ & g(\lambda) & g'(\lambda) & \dfrac{g''(\lambda)}{2!} & \cdots & \dfrac{g^{(r-2)}(\lambda)}{(r-2)!} \\ & & g(\lambda) & g'(\lambda) & \cdots & \dfrac{g^{(r-3)}(\lambda)}{(r-3)!} \\ & & & & \cdots & \cdot \\ & & & & \cdot\cdot & \cdot \\ & & & & \cdot & g'(\lambda) \\ & & & & & g(\lambda) \end{array} \tag{14}$$

where all the terms to the left of the main diagonal are 0, the coordinates in the first row are as indicated, and all those on a line parallel to the main diagonal are the same as the one where this line meets the first row.

If the characteristic function is the same as the reduced function, no two blocks of terms in (9) correspond to the same root and e_i, z_i are the principal idempotent and nilpotent elements of x corresponding to λ_i and (13) is the same as (4). This is not the case when the same root occurs in more than one of the blocks (10) and, when this is so, the a_i are not necessarily uniquely determined. For instance let

$$x = \begin{Vmatrix} \gamma & 0 & 0 \\ 0 & \gamma & 1 \\ 0 & 0 & \gamma \end{Vmatrix} = \gamma + e_{23}. \tag{15}$$

Here we have

$$a_1 = \gamma e_{11}, \qquad a_2 = \gamma(e_{22} + e_{33}) + e_{23},$$
$$e_1 = e_{11}, \qquad e_2 = e_{22} + e_{33}, \qquad z_1 = 0, \qquad z_2 = e_{23}.$$

But if

$$\begin{array}{lll} f_{11} = e_{11} - e_{13} & f_{12} = e_{12} & f_{13} = e_{13} \\ f_{21} = e_{21} - e_{23} & f_{22} = e_{22} & f_{23} = e_{23} \\ f_{31} = e_{31} - e_{13} + e_{11} - e_{33}, & f_{32} = e_{32} + e_{12}, & f_{33} = e_{33} + e_{13}, \end{array}$$

the f's form a set of matric units and

$$x = \gamma f_{11} + \gamma(f_{22} + f_{33}) + f_{23}$$

so that we might have chosen $f_1 = f_{11}$, $f_2 = f_{22} + f_{33} = e_{22} + e_{33} + e_{13}$ as idempotent elements in place of e_1 and e_2.

It should be carefully noted that f_1, f_2 are not commutative with e_1, e_2 and in consequence different determinations of a multiform function may not be commutative with each other. For instance, if x is the matrix given in (15) with $\gamma \neq 0$, and $\gamma^{\frac{1}{2}}$ is a particular determination of the square root of γ, we have already seen in §2.13 that determinations of $x^{\frac{1}{2}}$ are given by

$$\begin{aligned} u_1 &= \gamma^{\frac{1}{2}} e_{11} - \gamma^{\frac{1}{2}}(e_{22} + e_{33}) - e_{23}/2\gamma^{\frac{1}{2}} \\ u_2 &= \gamma^{\frac{1}{2}} f_{11} - \gamma^{\frac{1}{2}}(f_{22} + f_{33}) - f_{23}/2\gamma^{\frac{1}{2}} \\ &= \gamma^{\frac{1}{2}}(e_{11} - e_{13}) - \gamma^{\frac{1}{2}}(e_{22} + e_{33} + e_{13}) - e_{23}/2\gamma^{\frac{1}{2}} \\ &= u_1 - 2\gamma^{\frac{1}{2}} e_{13}, \end{aligned}$$

and these two values of $x^{\frac{1}{2}}$ are not commutative.

8.04 **Roots of 0 and 1.** The reduced equation of a nilpotent matrix of index m is $x^m = 0$ and this matrix can therefore be defined as a primitive mth root of 0; the index m cannot be greater than n and it exceeds 1 unless $x = 0$. The canonical form of x must contain at least one block of order $r_1 = m$, similar to (10) but with $\lambda_i = 0$, and a number of like blocks of orders, say, r_i ($i = 2, 3, \cdots$) where $r_i \leq r_1$ and $\sum_1 r_i = n$. This gives rise to a series of distinct types in number equal to the number of partitions of $n - m$ into parts no one of which exceeds m, and x is a primitive mth root if, and only if, it is similar to one of these types.

If x is a primitive mth root of 1, its reduced characteristic function is a factor of $\lambda^m - 1$ and hence x has simple elementary divisors. Let ϵ be a scalar primitive mth root of 1, and let $f_1, f_2, \cdots, f_s$ be idempotent matrices of ranks $r_1, r_2, \cdots$ for which $f_i f_j = 0$ $(i \neq j)$, $\Sigma f_i = 1$; for instance, if $\rho_i = \sum_{j=1}^{i-1} r_j$, we may set

$$f_i = \sum_{\rho_i+1}^{\rho_{i+1}} e_{jj} \qquad (i = 1, 2, \cdots, s; \rho_{s+1} = n).$$

The canonical form for x is then

$$\epsilon^{t_1} f_1 + \epsilon^{t_2} f_2 + \cdots + \epsilon^{t_s} f_s \tag{16}$$

where the exponents t_i are all different modulo m and at least one ϵ^{t_i}, say the first, is primitive. Any primitive mth root of 1 is then similar to a matrix of the form (16), and conversely.

8.05 The equation $y^m = x$; algebraic functions. Let $\lambda_1, \lambda_2, \cdots, \lambda_s$ be the distinct roots of x and $\mu_i = \lambda_i^{1/m}$ a particular determination of the mth root of λ_i for $i = 1, 2, \cdots, s$; then, if $y^m = x$, the roots of y are all of the form $\epsilon^{t_i}\mu_i$, where ϵ is a primitive scalar mth root of 1. Suppose that the roots of y are

$$\begin{array}{l} \mu_{11} = \mu_1, \quad \mu_{12} = \epsilon^{t_{12}}\mu_1, \quad \cdots, \quad \mu_{1r_1} = \epsilon^{t_{1r_1}}\mu_1, \\ \cdots\cdots\cdots\cdots\cdots\cdots\cdots\cdots\cdots\cdots \\ \mu_{s1} = \mu_s, \quad \mu_{s2} = \epsilon^{t_{s2}}\mu_s, \quad \cdots, \quad \mu_{sr_s} = \epsilon^{t_{sr_s}}\mu_s, \end{array} \tag{17}$$

and let a particular choice of the partial idempotent and nilpotent elements corresponding to μ_{ij} be f_{ijk} and h_{ijk} $(k = 1, 2, \cdots)$; also let the index of h_{ijk} be n_{ijk}. Then

$$y = \Sigma(\mu_{ij} f_{ijk} + h_{ijk}) \tag{18}$$

and hence

$$y^m = \Sigma(\mu_{ij} f_{ijk} + h_{ijk})^m = \Sigma(\lambda_i f_{ijk} + g_{ijk}) \tag{19}$$

where g_{ijk} is the nilpotent matrix

$$g_{ijk} = (\mu_{ij} f_{ijk} + h_{ijk})^m - \mu_{ij}^m f_{ijk}. \tag{20}$$

Further, if $\lambda_i \neq 0$, (20) can be solved for h_{ijk} as a polynomial in g_{ijk}; for we can write (20) in the form

$$g_{ijk} = \alpha_1 h_{ijk} + \alpha_2 h_{ijk}^2 + \cdots \tag{21}$$

and, since $\alpha_1 = m\mu_{ij}^{m-1} \neq 0$, the ordinary process for inverting a power series shows that we can satisfy (21) by a series of the form

$$h_{ijk} = \beta_1 g_{ijk} + \beta_2 g_{ijk}^2 + \cdots, \qquad (\beta_1 \neq 0), \tag{22}$$

there being here no question of convergence since any power series in a nilpotent matrix terminates. It follows from (21) and (22) that the indices of g_{ijk} and h_{ijk} are the same.

We shall now show that the matrices f_{ijk} and g_{ijk} form a set of partial idempotent and nilpotent elements of x provided always that x is not singular. If this were not so, then f_{ijk} must be the sum of two or more partial idempotent elements; for the sake of brevity we shall assume that it is the sum of two since the proof proceeds in exactly the same way if more components are taken. Let $f_{ijk} = d_1 + d_2$ where d_1 and d_2 are partial idempotent elements of x and let c_1, c_2 be the corresponding nilpotent elements; then

$$g_{ijk} = c_1 + c_2, \qquad c_1c_2 = 0 = c_2c_1.$$

Hence also $h_{ijk} = b_1 + b_2$, $b_1b_2 = 0 = b_2b_1$ where b_α is obtained by putting c_α for g_{ijk} in (22); and this is impossible since we assumed that f_{ijk} and h_{ijk} were partial idempotent and nilpotent elements of y. We have therefore the following theorem.

THEOREM 1. *If x is a non-singular matrix, any determination of $y = x^{1/m}$ can be obtained by expressing x in terms of partial idempotent and nilpotent elements, say $x = \Sigma(\lambda_i f_i + g_i)$ and putting*

$$y = \Sigma(\lambda_i f_i + g_i)^{1/m} = \Sigma\lambda_i^{1/m}(f_i + m^{-1}\lambda_i^{-1}g_i + \tfrac{1}{2}m^{-1}(m^{-1} - 1)\lambda_i^{-2}g^2 + \cdots).$$

Here the binomial series terminates and $\lambda_i^{1/m}$ is a determination of the mth root of λ_i which may be different for different terms of the summation if this root occurs with more than one partial element.

There is thus a two-sided multiplicity of mth roots of x; the $\lambda_i^{1/m}$ have m possible determinations in each term and also there is in certain cases an infinity of ways of choosing the set of partial elements. Since the canonical form is independent of the actual choice of the set of partial elements out of the possible sets, any choice of such a set can be derived from any other such set by transforming it by a matrix u; and since x itself is the same no matter what set of partial elements is chosen, we have $uxu^{-1} = x$, that is, u is commutative with x. It follows from the development given in §§7.01, 7.02 and 7.04 that a matrix u which is commutative with *every* partial idempotent element is a polynomial in x.

8.06 We must now consider the case in which x is singular and in doing so it is sufficient to discuss mth roots of a nilpotent matrix; for the principal idempotent element of x which corresponds to a root μ is the sum of those principal idempotent elements of y which correspond to those roots whose mth power is μ, so that the principal idempotent element corresponding to the root 0 is the same for both x and y. Let the elementary divisors of y be $\lambda^{m_1}, \lambda^{m_2}, \cdots, \lambda^{m_p}$; then

$$y = y_1 + y_2 + \cdots + y_p$$

where y_i is a nilpotent matrix of index m_i, and we may suppose the fundamental basis so chosen that the significant part of y_i is

$$\begin{matrix} 0 & 1 & & & \\ & 0 & 1 & & \\ & & \ddots & \ddots & \\ & & & 0 & 1 \\ & & & & 0 \end{matrix} \qquad (m_i \text{ rows and columns}). \tag{23}$$

To simplify the notation we shall consider for the moment only one part y_i and replace it by y and m_i by n so that $y^n = 0$ and

$$y = \sum_1^{n-1} e_i S e_{i+1}.$$

If we now form the mth power of y, then $y^m = 0$ if $m \geq n$ and if $m < n$

$$y^m = \sum_1^{n-m} e_i S e_{i+m}.$$

If we define r and k by

$$(r-1)m + k = n \leq rm \qquad (k > 0) \tag{24}$$

then $r \geq 2$ and

$$y^m e_i = 0,\ y^m e_{i+m} = e_i,\ y^m e_{i+2m} = e_{i+m},\ \cdots,\ y^m e_{i+(r-1)m} = e_{i+(r-2)m} \qquad (i = 1, 2, \cdots, k) \tag{25}$$

giving k chains of order r of invariant vectors, and similarly for $i = k + 1$, $k + 2, \cdots, m$, we have $m - k$ chains whose order is $r - 1$ since for these values of i the last equation in (25) is missing. If we set u and v for blocks of terms like (23) only with r and $r - 1$ rows and columns, respectively, then we can find a non-singular matrix P which permutes the rows and columns in y^m so that

$$P^{-1}y^mP = \left\| \begin{matrix} u & & & & & & \\ & u & & & & & \\ & & \ddots & & & & \\ & & & u & & & \\ & & & & v & & \\ & & & & & \ddots & \\ & & & & & & v \end{matrix} \right\| \qquad (k\ u\text{'s and } m - k\ v\text{'s}). \tag{26}$$

We are now in a position to consider the solution of $y^m = x$ where x is a nilpotent matrix of index r. In the elementary divisors of x suppose p_1 exponents equal r, p_2 equal $r - 1$, and in general p_j equal $r - j + 1$ $(j = 1, 2, \cdots, r)$; here the p's are integers equal to or greater than 0 such that

$$\sum_j (r - j + 1)p_j = n, \qquad p_1 \neq 0.$$

The maximum possible exponent for any elementary divisor of y is rm; let q_{ji} $(i = 1, 2, \cdots, m; j = 1, 2, \cdots, r)$ be the number of exponents which equal

$$(r - j + 1)m - i + 1 = (r - j)m + (m - i + 1).$$

Forming y^m and using (24) and the results of (25) we then see that

$$\begin{aligned} q_{j-1,2} + 2q_{j-1,3} + \cdots + (m-1)q_{j-1,m} + mq_{j1} + (m-1)q_{j2} \\ + \cdots + q_{jm} = p_j \qquad (j = 1, 2, \cdots, r; q_{0i} = 0) \end{aligned} \tag{27}$$

$$\sum_{i,j} [(r - j + 1)m - i + 1]\, q_{ji} = n.$$

These relations form a set of Diophantine equations for the q's. When a set of q's have been found, we can find the matrix P (cf. (26)) for each part of y^m and then set $y = \Sigma R^{-1}y_iR$ where R has the form ΣP_iQ_i, Q_i being commutative with $P_i^{-1}y_i^mP_i$ and so chosen that R is not singular.

8.07 **The exponential and logarithmic functions.** The function exp $y = \epsilon^y$ has already been defined in §8.02 by the series

$$\epsilon^y = 1 + \sum_1^\infty y^m/m!$$

or in §8.03 in terms of the partial units of y. Let the distinct roots of y be $\mu_1, \mu_2, \cdots, \mu_s$ and let a choice of the partial idempotent and nilpotent elements corresponding to μ_i be f_{ij}, h_{ij} $(j = 1, 2, \cdots, k_i)$ so that

$$f_i = \sum_{j=1}^{k_i} f_{ij}, \qquad h_i = \sum_{j=1}^{k_i} h_{ij} \qquad (i = 1, 2, \cdots, s) \tag{28}$$

are the principal idempotent and nilpotent elements of y. If we set $x = \epsilon^y$, we have

$$\begin{aligned} x = \epsilon^y &= \sum_i \sum_j \left[\epsilon^{\mu_i}f_{ij} + h_{ij} + \frac{h_{ij}^2}{2!} + \cdots + \frac{h_{ij}^{\nu_{ij}-1}}{(\nu_{ij} - 1)!}\right] \\ &= \sum_i \sum_j (\epsilon^{\mu_i}f_{ij} + g_{ij}) \end{aligned} \tag{29}$$

where ν_{ij} is the index of h_{ij} and

$$g_{ij} = h_{ij} + \frac{h_{ij}^2}{2} + \cdots + h_{ij}^{\nu_{ij}-1}/(\nu_{ij} - 1)!. \tag{30}$$

The index of g_{ij} is clearly ν_{ij}. Solving (30) for h_{ij}, we have the formal solution $h_{ij} = \log(1 + g_{ij})$ and on using this or inverting the power series in (30) we get

$$(31) \qquad h_{ij} = g_{ij} - \tfrac{1}{2}g_{ij}^2 + \cdots + (-1)^{\nu_{ij}} g_{ij}^{\nu_{ij}-1}/(\nu_{ij} - 1).$$

As in §8.05 it follows that f_{ij}, g_{ij} form a set of partial elements for x and, when x is given so that $y = \log x$, the method there used gives the following theorem.

Theorem 2. *If x is a non-singular matrix whose distinct roots are $\lambda_1, \lambda_2, \cdots, \lambda_r$, and if* $\log \lambda_1, \log \lambda_2, \cdots, \log \lambda_r$ *are particular determinations of the logarithms of these roots, then the general determination of* $\log x$ *is found as follows. Take any set of partial elements of x, say f_{ij}, g_{ij} $(i = 1, 2, \cdots, r; j = 1, 2, \cdots, k_i)$ where f_{ij}, g_{ij} correspond to λ_i and the index of g_{ij} is ν_{ij}, let h_{ij} be the nilpotent matrix defined by* (31), *and let k_{ij} be any integers, then*

$$(32) \qquad \log x = \sum_i \sum_j [(\log \lambda_i + k_{ij}\pi\sqrt{-1})f_{ij} + h_{ij}].$$

The discussion of the relation between different determinations of $\log x$ is practically the same as for $x^{1/m}$ and need not be repeated.

If f_i and h_i are defined by (28), a particular determination of $\log x$ is given by

$$(33) \qquad \operatorname{Log} x = \sum_i [(\log \lambda_i + k_i\pi\sqrt{-1})f_i + h_i].$$

This form of $\log x$ has the same principal elements as x provided $\log \lambda_i + k_i \neq \log \lambda_j + k_j$ for any $i \neq j$, and even when this condition is not satisfied, it is convenient to refer to (33) as a *principal* determination of $\log x$. This determination is the one given by the series (cf. §8.02 (8))

$$(34) \qquad \log x = (x - 1) - \tfrac{1}{2}(x - 1)^2 + \tfrac{1}{3}(x - 1)^3 - \cdots$$

provided each k_i is 0 and the principal determination of $\log \lambda_i$ is used. The series converges only when the roots of $x - 1$ are all less than 1 in absolute value.

8.08 The canonical form of a matrix in a given field. If the coefficients of a matrix are restricted to lie in a given field of rationality, the canonical form used in the preceding sections requires some modification. The definition of the invariant factors is rational as are also the theorems regarding similar matrices which were derived from them in Chapter 3; and hence if X and x are rational matrices which have the same invariant factors there exists a rational matrix P for which $P^{-1}xP = X$. The definition of elementary divisors requires only the natural alteration of substituting powers of irreducible polynomials for $(\lambda - \lambda_i)^{\nu_{ij}}$.

Let

$$\alpha(\lambda) = \lambda^m + a_1\lambda^{m-1} + \cdots + a_m$$

be a scalar polynomial in a field F which is irreducible in F; then the matrix of order m defined by

$$x_a = \begin{matrix} -a_1 & -a_2 & \cdots & -a_{m-1} & -a_m \\ 1 & 0 & \cdots & 0 & 0 \\ 0 & 1 & \cdots & 0 & 0 \\ \cdot & \cdot & \cdots & \cdot & \cdot \\ \cdot & \cdot & \cdots & \cdot & \cdot \\ 0 & 0 & \cdots & 1 & 0 \end{matrix} \tag{35}$$

has $\alpha(\lambda)$ as its characteristic function; since $\alpha(\lambda)$ is irreducible in F, it follows immediately that x_a is an irreducible matrix in F and that $\alpha(\lambda)$ is also the reduced characteristic function. It is easily seen that the invariant factors of $\lambda - x$ are given by $m - 1$ units followed by $\alpha(\lambda)$.

Again, if we consider

$$x_\alpha = \begin{matrix} x_a & 1_m & & & \\ & x_a & 1_m & & \\ & & \cdot & \cdot & \\ & & & \cdot & \cdot \\ & & & \cdot & 1_m \\ & & & & x_a \end{matrix} \qquad (r \text{ rows and columns}) \tag{36}$$

which is a matrix of order rm, we see as in §8.03 (14) that, if $g(\lambda)$ is a scalar polynomial in λ,

$$g(x_\alpha) = \begin{matrix} g(x_a) & g'(x_a) & \cdots & g^{(r-1)}(x_a)/(r-1)! \\ & g(x_a) & \cdots & g^{(r-2)}(x_a)/(r-2)! \\ & & \cdots & \cdot \\ & & \cdot\cdot & \cdot \\ & & \cdot & g'(x_a) \\ & & & g(x_a). \end{matrix}$$

It follows that, if $g(x_\alpha) = 0$, we must have $g^{(r-1)}(x_a) = 0$ and therefore $\alpha(\lambda)$ is a factor of $g^{(r-1)}(\lambda)$ so that $[\alpha(\lambda)]^r$ is a factor of $g(\lambda)$. But if we put $g(\lambda) = [\alpha(\lambda)]^r$ the first $(r - 1)$ derivatives of $g(\lambda)$ have $\alpha(\lambda)$ as a factor and so vanish when λ is replaced by x_a; hence $g(x_\alpha) = 0$. It follows that the reduced characteristic function of x_α is $[\alpha(\lambda)]^r$ and, since the degree of this polynomial equals the order rm of x_α, it is also the characteristic function so that the invariant factors of x_α are given by 1 repeated $rm - 1$ times followed by $[\alpha(\lambda)]^r$. The argument used in §3.06 then gives the following theorem.

THEOREM 3. *Let $\alpha_1(\lambda), \alpha_2(\lambda), \cdots, \alpha_k(\lambda)$ be polynomials, not necessarily distinct, which are rational and irreducible in a field F and whose degrees are $m_1, m_2, \cdots, m_k$ respectively; and let $r_1, r_2, \cdots, r_k$ be any positive integers such that*

$\sum_{1}^{k} r_i m_i = n$; *then, if* x_{α_i} *is the matrix of order* $r_i m_i$ *formed from* $\alpha_i(\lambda)$ *in the same way as* x_α *in* (36) *is formed from* $\alpha(\lambda)$, *the matrix of order* n *defined by*

$$x = \begin{matrix} x_{\alpha_1} & & & & \\ & x_{\alpha_2} & & & \\ & & \cdot & & \\ & & & \cdot & \\ & & & & \cdot \\ & & & & & x_{\alpha_k} \end{matrix} \tag{37}$$

has $\alpha_1(\lambda), \alpha_2(\lambda), \cdots, \alpha_k(\lambda)$ *as its elementary divisors in* F.

If X is any matrix with the same elementary divisors as x, it follows from §3.04 Theorem 5 that we can find a rational nonsingular matrix P such that $PXP^{-1} = x$. We may therefore take (37) as a canonical form for a matrix in the given field F.

8.09 **The absolute value of a matrix.** The absolute value of a matrix $a = \| a_{pq} \|$ is most conveniently defined as

$$\mathbf{|} a \mathbf{|} = \left(\sum_{p,q=1}^{n} a_{pq}\bar{a}_{pq} \right)^{\frac{1}{2}} \tag{38}$$

where the heavy bars are used to distinguish between the absolute value and the determinant $| a |$. It must be carefully noted that the absolute value of a scalar matrix λ is not the same as the ordinary absolute value or modulus of λ, the relation between them being

$$\mathbf{|} \lambda \mathbf{|} = n^{\frac{1}{2}} \operatorname{mod} \lambda. \tag{39}$$

It follows immediately from (37) that

$$\operatorname{mod} (\mathbf{|} a \mathbf{|} - \mathbf{|} b \mathbf{|}) \leq \mathbf{|} a + b \mathbf{|} \leq \mathbf{|} a \mathbf{|} + \mathbf{|} b \mathbf{|}; \tag{40}$$

and from

$$\sum_{p,r} a_{pr}\bar{a}_{pr} \sum_{qs} b_{sq}\bar{b}_{sq} = \sum_{p,q} \left[\tfrac{1}{2} \sum_{r,s} (a_{pr}\bar{b}_{sq} - a_{ps}\bar{b}_{rq})(\bar{a}_{pr}b_{sq} - \bar{a}_{ps}b_{rq}) + \sum_{r} a_{pr}b_{rq} \sum_{s} \bar{a}_{ps}\bar{b}_{sq} \right],$$

we have

$$\mathbf{|} ab \mathbf{|} \leq \mathbf{|} a \mathbf{|}\mathbf{|} b \mathbf{|}. \tag{41}$$

Since the trace of $a\bar{a}'$ is $\Sigma a_{pq}\bar{a}_{pq}$, the absolute value of a might also have been defined by

$$\mathbf{|} a \mathbf{|}^2 = \operatorname{tr} a\bar{a}' = \operatorname{tr} \bar{a}'a. \tag{42}$$

From this we see immediately that, if a is unitary, that is, $a\bar{a}' = 1$, then

$$|a| = n^{\frac{1}{2}}, \qquad |ab| = |b| \tag{43}$$

where b is any matrix.

No matter what matrix a is, $a\bar{a}'$ is a positive hermitian matrix, semi-definite or definite according as a is or is not singular; the roots $g_1, g_2, \cdots, g_n$ of $a\bar{a}'$ are therefore real and not negative. If we set

$$p_r = \Sigma g_1 g_2 \cdots g_r, \qquad p \equiv p_n = (\mathrm{mod}\, |a|)^2, \qquad s \equiv p_1 = |a|^2 = \Sigma g_i,$$

then

$$r^r p_r \leq \Sigma(g_1 + g_2 + \cdots + g_r)^r \leq \binom{n-1}{r-1} s^r, \qquad p_r \geq \binom{n}{r} p^{\binom{n-1}{r-1}/\binom{n}{r}} = \binom{n}{r} p^{r/n},$$

whence[1]

$$\binom{n}{r} p^{r/n} \leq p_r \leq r^{-r} \binom{n-1}{r-1} s^r. \tag{44}$$

If $C^r(a)$ is the rth supplementary compound of a (cf. §5.03), and $\alpha = p^{\frac{1}{2}}$ is put for mod $|a|$, then $p_r = |C^r(a)|^2$ and we may write for (44)

$$\binom{n}{r} \alpha^{2r/n} \leq |C^r(a)|^2 \leq r^{-r} \binom{n-1}{r-1} |a|^{2r}$$

and, since $(a')^{-1} = C^{n-1}(a)/|a|$, we have

$$n^{\frac{1}{2}} \alpha^{-1/n} \leq |\alpha^{-1}| \leq (n-1)^{-\frac{n-2}{2}} |a|^{n-1}/\alpha \tag{45}$$

provided $|a| \neq 0$. This inequality enables us to deal with expressions involving negative powers of a.

Since $a^{-1} = \exp(-\log a)$, we also have

$$|a^{-1}b| = |(1 - \log a + \tfrac{1}{2!}(\log a)^2 - \cdots)b| \leq (1 + |\log a| + \tfrac{1}{2!}|\log a|^2 + \cdots)|b|$$

and therefore

$$|a^{-1}b| \leq \epsilon^{|\log a|} |b|. \tag{46}$$

Putting $b = 1$ we also have

$$|a^{-1}| \leq n^{\frac{1}{2}} - 1 + \epsilon^{|\log a|} \quad \text{and} \quad |a^{-1}| \leq n^{\frac{1}{2}} \epsilon^{|\log a|}. \tag{47}$$

It is also sometimes convenient to note as a consequence of (41) with $b = a^{-1}$ that

$$|a^{-1}| \geq n^{\frac{1}{2}} |a|^{-1}. \tag{48}$$

[1] If $r = n$, (44) gives Hadamard's expression for the maximum value of mod $|a|$.

8.10 **Infinite products.** As an illustration of the use of the preceding section we shall now investigate briefly the convergence of an infinite product. Let

$$(49)\qquad P_m = (1 + a_1)(1 + a_2) \cdots (1 + a_m)$$

$$Q_m = (1 + |a_1|)(1 + |a_2|) \cdots (1 + |a_m|),$$

then, if c is an arbitrary matrix,

$$(50)\qquad |P_m - 1| \leq Q_m - 1 < \epsilon^{\Sigma |a_i|} - 1,$$

$$(51)\qquad |P_m c| \leq \epsilon^{\Sigma |a_i|} |c|,$$

$$(52)\qquad |P_m - P_k| \leq Q_m - Q_k \leq \sum_k^m |a_i| \epsilon^{\sum_1^m |a_i|}.$$

For on expanding P_m we have

$$P_m = 1 + \sum_p a_p + \sum_{p<q} a_p a_q + \sum_{p<q<r} a_p a_q a_r + \cdots$$

therefore

$$|P_m - 1| \leq \sum_p |a_p| + \sum_{p<q} |a_p||a_q| + \cdots = Q_m - 1$$

$$|P_m c| \leq (1 + \sum_p |a_p| + \sum_{p<q} |a_p||a_q| + \cdots)|c| = Q_m |c| < \epsilon^{\Sigma |a_i|} |c|.$$

The proof of (52) follows in the same manner.

Hence P_m converges when Q_m does, for which it is sufficient that Σa_i is absolutely convergent in the sense that $\Sigma |a_i|$ converges.

8.11 **The absolute value of a tensor.** If w is a tensor of grade r, we define the absolute value of w by[2]

$$(53)\qquad \text{mod}\, w = (r! S\bar{w}w)^{\frac{1}{2}}.$$

We shall for the most part consider only vectors of grade 1 as the extension to tensors of higher grade is usually immediate.

If x and y are any vectors, we derive from (53)

$$(54)\qquad \text{mod}\,(x + y) \leq \text{mod}\, x + \text{mod}\, y, \qquad \text{mod}\, Sxy \leq \text{mod}\, x \,\text{mod}\, y.$$

If A is a matrix,

$$(\text{mod}\, Ax)^2 = S\bar{A}\bar{x}Ax = S\bar{x}\bar{A}'Ax.$$

By §6.02

$$\bar{A}'A = \Sigma g_i \bar{a}_i S a_i$$

[2] The $r!$ enters here only because of the numerical factor introduced in defining Suv (cf. §5.16).

where the g's are real and not negative, and $S\bar{a}_i a_j = \delta_{ij}$ so that mod $a_i = 1$; hence

$$\begin{aligned} S\bar{x}\bar{A}'Ax &= \Sigma g_i S\bar{x}\bar{a}_i Sa_i x = \Sigma g_i \,(\text{mod } Sa_i x)^2 \\ &\leq \Sigma g_i \,(\text{mod } a_i \text{ mod } x)^2 = (\Sigma g_i)(\text{mod } x)^2 \\ &= |\, A\,|^2 \,(\text{mod } x)^2, \end{aligned}$$

or

$$\text{mod } Ax \leq |\, A\,| \text{ mod } x. \tag{55}$$

From (54) we then have

$$\text{mod } SyAx \leq |\, A\,| \text{ mod } x \text{ mod } y. \tag{56}$$

8.12 **Matric functions of a scalar variable.** If the coordinates of a matrix $a(t) = \|\, a_{pq}(t)\,\|$ are functions of a scalar variable t, the matrix itself is called a matric function of t. The derivative, when it exists, is defined as

$$\frac{da}{dt} = \lim_{h\to 0} \frac{a(t+h) - a(t)}{h} = \left\| \frac{da_{pq}}{dt} \right\| \tag{57}$$

h being a scalar. The fundamental rules of differentiation are

$$\frac{d(a+b)}{dt} = \frac{da}{dt} + \frac{db}{dt}, \qquad \frac{d(ab)}{dt} = \frac{da}{dt}\,b + a\,\frac{db}{dt}, \qquad \frac{da'}{dt} = \left(\frac{da}{dt}\right)'$$

to which we may add, when $|\, a \,| \neq 0$,

$$\frac{da^{-1}}{dt} = -a^{-1}\,\frac{da}{dt}\,a^{-1}. \tag{58}$$

Other examples are

$$\frac{da^2}{dt} = \frac{da}{dt}\,a + a\,\frac{da}{dt}, \qquad \frac{da^3}{dt} = \frac{da}{dt}\,a^2 + a\,\frac{da}{dt}\,a + a^2\,\frac{da}{dt}$$

and in general, if m is any positive integer,

$$\frac{da^m}{dt} = \begin{Bmatrix} a & da/dt \\ m-1 & 1 \end{Bmatrix}. \tag{59}$$

Under the usual conditions each of the coordinates of $a(t)$ is expansible as a Taylor series and this is therefore also true of $a(t)$. If $f(t)$ is a scalar function, $f(a)$ may or may not have a meaning. For instance, if $f(t)$ can be expanded in a power series which converges for mod $t < \alpha$, then the same power series[3] in

[3] If $g(t) = \Sigma u_n t^n$, u_n scalar, the series intended here is $\Sigma u_n a^n$. Other definitions are possible, e.g., if we set

$$G(a) = \sum_n u_n \sum_i c_{i0}^{(n)}\, ac_{i1}^{(n)}\, ac_{i2}^{(n)} \cdots c_{i,\,n-1}^{(n)} ac_{in}^{(n)}$$

where $\sum_i c_{i0}^{(n)}\, c_{i1}^{(n)} \cdots c_{in}^{(n)} = 1$, we still have $G(t) = g(t)$.

a converges when $|a| < \alpha$; but if $f(t)$ is defined by a Fourier series which is not differentiable, $f(a)$ will not have a meaning when the elementary divisors of a are not simple, as is seen immediately on referring to the form of §8.03 (14). If $f(a)$ and $f'(a)$ have a meaning and if da/dt is commutative with a, we have $df(a)/dt = f'(a)da/dt$. For instance, if x is a constant matrix and $a = t - x$, then

$$\frac{d \log (t - x)}{dt} = \frac{1}{t - x} \tag{60}$$

as is also easily proved directly.

The integral of $a(t)$ is defined as follows. If C is a regular contour in the t-plane, we shall set

$$\int_C a(t)dt = \left\| \int_C a_{pq}(t)dt \right\| \tag{61}$$

or if $t_1, t_2, \cdots$ is a series of points on C and t'_i a point on the arc (t_i, t_{i+1}), and if the number of points is increased indefinitely in such a way that mod $(t_{i+1} - t_i)$ approaches 0 for every interval, then

$$\int_C a(t)dt = \lim \Sigma a(t'_i)(t_{i+1} - t_i). \tag{62}$$

The conditions for the existence of this limit are exactly the same as in the scalar theory.

If M is the least upper bound of $|a|$ on C and L is the length of C, it follows in the usual manner that

$$\left| \int_C a(t)dt \right| \leq \int |a(t)| \operatorname{mod} dt \leq ML. \tag{63}$$

As an illustration of these definitions we shall now employ contour integration to prove some of our earlier results. If x is an arbitrary constant matrix and C a circle with center $t = 0$ and radius greater than $|x|$, then all the roots of x lie inside C and on C the series

$$\frac{1}{t - x} = \frac{1}{t} + \frac{x}{t^2} + \frac{x^2}{t^3} + \cdots$$

is uniformly convergent. Hence

$$\frac{1}{2\pi i} \int_C \frac{dt}{t - x} = \frac{1}{2\pi i} \sum_0^\infty x^n \int_C \frac{dt}{t^{n+1}} = 1 \tag{64}$$

a result which may also be derived from the definition of log $(t - x)$ in §8.07 and

$$\int_C \frac{dt}{t - x} = [\log (t - x)]_C.$$

We then have

$$\frac{1}{2\pi i} \int_C \frac{tdt}{t - x} = \frac{1}{2\pi i} \int_C \left(1 + \frac{x}{t - x}\right) dt = x$$

and in general, if $g(t)$ is a scalar function of t which is analytic in a region enclosing C,

(65) $$g(x) = \frac{1}{2\pi i} \int_C \frac{g(t)}{t - x}\, dt.$$

Suppose now that $| t - x | = \theta(t)\varphi(t)$, adj $(t - x) = \theta(t)a(t)$ where $\theta(t)$ is the highest common factor of $| t - x |$ and the coordinates of adj $(t - x)$. We then have

(66) $$g(x) = \frac{1}{2\pi i} \int_C \frac{g(t)}{\varphi(t)}\, a(t)dt$$

and under the given conditions this vanishes if, and only if, $g(t)/\varphi(t)$ has no singularities inside C, that is, if φ is a factor of g. We have therefore the theorem of §2.05 that $\varphi(t)$ is the reduced characteristic function of x and that $g(x) = 0$ only when $\varphi(t)$ is a factor of $g(t)$.

Since $a(t) \equiv \varphi(t)/(t - x)$ is a polynomial in x with scalar coefficients and with degree 1 less than the degree of $\varphi(t)$, say

$$a(t) = \alpha_1 x^{m-1} + \alpha_2 x^{m-2} + \cdots + \alpha_m,$$

equation (66) shows that $g(x)$ can be expressed as a polynomial in x, namely,

(67) $$g(x) = \frac{1}{2\pi i} \Sigma x^{m-s} \int_C \frac{g(t)\alpha_s(t)}{\varphi(t)}\, dt.$$

We may also note that (66) leads to the interpolation formula §8.01 (4) if the integral is expanded in terms of the residues at the zeros of $\varphi(t)$.

All of these results can be extended to unilateral series in x with matric coefficients if care is taken to use $g(t)\,(t - x)^{-1}$ or $(t - x)^{-1}\,g(t)$ according as dextro- or laevo-lateral series are desired.

8.13 Functions of a variable vector. Before considering functions of a variable matrix, we shall consider briefly those of a variable vector; for more extended and systematic treatments the reader is referred to treatises on vector and tensor analysis.

The differential of a function of a variable in any non-commutative algebra was defined by Hamilton as follows. Let $f(x)$ be a function of a variable x, dx a variable independent of x and t a scalar variable; then

$$df(x) = \lim_{t\to 0} \frac{f(x + tdx) - f(x)}{t}. \tag{68}$$

We shall assume tacitly hereafter that this limit exists for all the functions we shall consider.

An immediate consequence of (68) is that $df(x)$ is linear and homogeneous in dx. Hence, if $x = \Sigma\xi_i e_i$, $dx = \Sigma d\xi_i e_i$, then

$$df(x) = \lim_{t\to 0} \frac{f(\Sigma(\xi_i + td\xi_i)e_i) - f(\Sigma\xi_i e_i)}{t} = \Sigma \frac{\partial f}{\partial \xi_i} d\xi_i.$$

This leads to Hamilton's differential operator

$$\nabla = \Sigma e_i \frac{\partial}{\partial \xi_i} \tag{69}$$

in terms of which we may write (68) in the form

$$df(x) = (Sdx\nabla)f(x). \tag{70}$$

In using this operator it is frequently convenient to place it after its operand and, when this is done, some artifice is necessary to indicate the connection between them. This is done by attaching the same subscript to both; the method of doing this will be clear from the following examples in which $a = \Sigma\alpha_i e_i$, $b = \Sigma\beta_i e_i$ are vectors and $A = \| a_{ij} \|$ is a matrix.

$$\nabla a = \Sigma \frac{\partial \alpha_j}{\partial \xi_i} e_i e_j, \qquad a_\alpha \nabla_\alpha = \Sigma \frac{\partial \alpha_i}{\partial \xi_j} e_i e_j,$$

$$\nabla Sab = \nabla_\alpha S a_\alpha b + \nabla_\alpha S a b_\alpha = \sum_i \left(\sum_j \left(\frac{\partial \alpha_j}{\partial \xi_i} \beta_j + \frac{\partial \beta_j}{\partial \xi_i} \alpha_j \right) \right) e_i,$$

$$S\nabla_\alpha \nabla_\beta S a_\alpha b_\beta = \sum_{i,j} \frac{\partial \alpha_i}{\partial \xi_j} \frac{\partial \beta_i}{\partial \xi_j},$$

$$A_\alpha \nabla_\alpha = \sum_i \left(\sum_j \frac{\partial a_{ij}}{\partial \xi_j} \right) e_i, \qquad \nabla_\alpha A_\alpha = \Sigma \frac{\partial a_{jk}}{\partial \xi_i} e_i e_j S e_k(\ \ ),$$

$$S a_\alpha A \nabla_\alpha = \sum_{i\,j} a_{ij} \frac{\partial \alpha_i}{\partial \xi_j}, \qquad S a A_\alpha \nabla_\alpha = \sum_{i,j} \alpha_i \frac{\partial a_{ij}}{\partial \xi_j}.$$

We can now consider the effect of a change of variable from x to[4] $\bar{x}$. Let $\bar{x} = \Sigma\bar{\xi}_i e_i$, $\bar{\nabla} = \Sigma e_i \partial/\partial\bar{\xi}_i$; then

$$d\bar{x} = \bar{x}_\alpha S \nabla_\alpha dx = J dx \tag{71}$$

[4] Here $\bar{x}$ denotes a new variable and not the conjugate imaginary. Instead of considering a change of variable, we may regard $\bar{x}$ as a vector function of x.

where

$$J = \bar{x}_\alpha S\nabla_\alpha = \| \partial\bar{\xi}_i/\partial\xi_j \| \tag{72}$$

is the Jacobian matrix of the transformation. Similarly

$$dx = x_\alpha S\bar{\nabla}_\alpha d\bar{x} = \bar{J}d\bar{x}.$$

Hence

$$J\bar{J} = 1. \tag{73}$$

Again, since $Sdx\nabla = Sd\bar{x}\bar{\nabla} = SdxJ'\bar{\nabla}$, hence

$$\nabla = J'\bar{\nabla}, \qquad \bar{\nabla} = (J')^{-1}\nabla = \bar{J}'\nabla. \tag{74}$$

From (70) and (72) we see that the differentials of J and J' are given by

$$\begin{aligned} dJ &= Sdx\nabla_\alpha \cdot J_\alpha, &\quad \text{and} \quad dJ &= d\bar{x}_\alpha S\nabla_\alpha = J_\alpha dx S\nabla_\alpha \\ dJ' &= Sdx\nabla_\alpha \cdot J'_\alpha, & dJ' &= \nabla_\alpha SJ_\alpha dx = \nabla_\alpha SdxJ'_\alpha. \end{aligned} \tag{75}$$

This leads to the notion of contravariant and covariant vectors. If u is a vector function of x and $\bar{u}$ the corresponding[5] function after the change of variable, u is called contravariant if

$$\bar{u} = Ju, \tag{76}$$

and covariant when

$$\bar{u} = \bar{J}'u, \qquad u = J'\bar{u}. \tag{77}$$

If d_1, d_2 denote two independent variations so that $d_1(d_2x) = d_2(d_1x)$, then

$$\begin{aligned} d_2d_1\bar{x} &= d_2(Jd_1x) = d_2Jd_1x + Jd_1d_2x \\ &= J_\alpha d_2xS\nabla_\alpha d_1x + Jd_1d_2x. \end{aligned} \tag{78}$$

Hence second differentials are neither contra- nor co-variant.

If A is a matrix whose coordinates are functions of x, the bilinear differential form Sd_1xAd_2x when transformed becomes

$$Sd_1\bar{x}\bar{A}d_2\bar{x} = Sd_1xJ'\bar{A}Jd_2x$$

so that, if this form is invariant, that is, $Sd_1\bar{x}\bar{A}d_2\bar{x} = Sd_1xAd_2x$, we must have

$$A = J'\bar{A}J, \qquad A' = J'\bar{A}'J, \qquad Adx = J'\bar{A}d\bar{x}. \tag{79}$$

[5] As will be seen below, this does not necessarily mean merely the result of substituting $\bar{x}$ for x in the coordinates of u.

Hence when $\bar{A}$ is defined in this manner, Adx is a covariant vector. If by analogy with (78) we form a second differential of this vector and of $A'dx$, we get, using $d\bar{x} = Jdx$,

$$d_2(Ad_1x) = d_2J'\bar{A}d_1\bar{x} + J'd_2(\bar{A}d_1\bar{x})$$
$$d_1(A'd_2x) = d_1J'\bar{A}'d_2\bar{x} + J'd_1(\bar{A}'d_2\bar{x}).$$

From (75) $d_2J' = \nabla_\alpha SJ_\alpha d_2x$, $d_1J' = \nabla_\alpha Sd_1xJ'_\alpha$; hence after a simple reduction

$$\begin{aligned} d_2(Ad_1x) + d_1(A'd_2x) &= \nabla_\alpha Sd_1x(J'\bar{A}'J_\alpha + J'_\alpha\bar{A}'J)d_2x + J'(d_2(\bar{A}d_1\bar{x}) + d_1(\bar{A}'d_2\bar{x})) \\ &= \nabla_\alpha Sd_1x(A'_\alpha - J'\bar{A}'_\alpha J)d_2x + J'(d_2(\bar{A}d_1\bar{x}) + d_1(\bar{A}'d_2\bar{x})) \end{aligned}$$

which may be written

$$\begin{aligned} a &\equiv d_2(Ad_1x) + d_1(A'd_2x) - \nabla_\alpha Sd_1xA'_\alpha d_2x \\ &= J'(d_2(\bar{A}d_1\bar{x}) + d_1(\bar{A}'d_2\bar{x}) - \nabla_\alpha Sd_1\bar{x}\bar{A}'_\alpha d_2\bar{x}) \\ &= J'\bar{a} \end{aligned} \tag{80}$$

so that a is a covariant vector. This vector may also be written

$$a = d_2Ad_1x + d_1A'd_2x - \nabla_\alpha Sd_1xA'_\alpha d_2x + (A + A')d_1d_2x.$$

Using a notation suggested by the Christoffel symbols we now write

$$\begin{aligned} [A; d_1x, d_2x] &= \tfrac{1}{2}(d_2Ad_1x + d_1A'd_2x - \nabla_\alpha Sd_1xA'_\alpha d_2x) \\ &= \sum_{i\,j\,k} \tfrac{1}{2}\left(\frac{\partial a_{kj}}{\partial \xi_i} + \frac{\partial a_{ki}}{\partial \xi_j} - \frac{\partial a_{ji}}{\partial \xi_k}\right) d_1\xi_i d_2\xi_j e_k \end{aligned} \tag{81}$$

$$\begin{aligned} \begin{Bmatrix} A \\ d_1x,\ d_2x \end{Bmatrix} &= (A + A')^{-1}(d_2Ad_1x + d_1A'd_2x - \nabla_\alpha Sd_1xA'_\alpha d_2x) \\ &= 2(A + A')^{-1}[A; d_1x, d_2x] \end{aligned} \tag{82}$$

provided that $|A + A'| \neq 0$. If we now set

$$b = \begin{Bmatrix} A \\ d_1x,\ d_2x \end{Bmatrix} + d_1d_2x$$

and use the relation $(A + A')^{-1}J' = J^{-1}(\bar{A} + \bar{A}')^{-1}$, we have from (80)

$$\bar{b} = Jb \tag{83}$$

so that b is contravariant.

If we set $A = \frac{1}{2}(A + A') + \frac{1}{2}(A - A') = B + C$, we get from (81) and (80)

$$\begin{aligned} [A; d_1x, d_2x] &= [B; d_1x, d_2x] + [C; d_1x, d_2x] \\ [B; d_1x, d_2x] + Bd_1d_2x &= J'([\bar{B}; d_1\bar{x}, d_2\bar{x}] + \bar{B}d_1d_2\bar{x}) \\ [C; d_1x, d_2x] &= J'[\bar{C}; d_1\bar{x}, d_2\bar{x}]. \end{aligned} \tag{84}$$

We shall require two transverses of the Christoffel matrices; these are defined by

$$Sb[A; a, c] = Sc[A; a, b]', \qquad Sb\begin{Bmatrix} A \\ a, c \end{Bmatrix} = Sc\begin{Bmatrix} A \\ a, b \end{Bmatrix}'$$

whence

$$(85) \qquad \begin{aligned} 2[A; a, b]' &= \nabla_\alpha SaA'_\alpha b + A_\alpha bSa\nabla_\alpha - A_\alpha aSb\nabla_\alpha \\ \begin{Bmatrix} A \\ a, b \end{Bmatrix}' &= [A; a, (A + A')^{-1}b]'. \end{aligned}$$

To illustrate partial differentiation we shall consider functions which depend not only on x but also on a contravariant variable vector $u = \Sigma\omega_i e_i$. Since $\bar{u} = Ju$ and $J = \| \partial\bar{\xi}_i/\partial\xi_j \|$ is independent of u, we have

$$\frac{\partial}{\partial\omega_i} = \sum \frac{\partial\bar{\omega}_j}{\partial\omega_i}\frac{\partial}{\partial\bar{\omega}_j} = \sum \frac{\partial\bar{\xi}_j}{\partial\xi_i}\frac{\partial}{\partial\bar{\omega}_j}$$

$$\frac{\partial}{\partial\xi_i} = \sum \frac{\partial\bar{\xi}_j}{\partial\xi_i}\frac{\partial}{\partial\bar{\xi}_j} + \sum \frac{\partial\bar{\omega}_j}{\partial\xi_i}\frac{\partial}{\partial\bar{\omega}_j}.$$

Hence, if $\nabla' = \Sigma e_i\partial/\partial\omega_i$, $\bar{\nabla}' = \Sigma e_i\partial/\partial\bar{\omega}_i$, then

$$(86) \qquad \nabla' = J'\bar{\nabla}', \qquad \nabla = J'\bar{\nabla} + d_uJ'\cdot\bar{\nabla}'$$

where $d_uJ' \equiv Su\nabla_\alpha\cdot J'_\alpha$. Here ∇' is covariant but ∇ is neither covariant nor contravariant, which means that formulae dependent on it will not usually be invariant in form under a change of variable. This difficulty is avoided as follows. If we combine (83) with (78) and replace d_1x, d_2x by contravariant vectors a, b, then

$$(87) \qquad d_bJa = d_aJb = J_\alpha bS\nabla_\alpha a = J\begin{Bmatrix} A \\ a, b \end{Bmatrix} - \begin{Bmatrix} \bar{A} \\ \bar{a}, \bar{b} \end{Bmatrix}.$$

Hence

$$(87') \qquad d_aJ'b = \begin{Bmatrix} A \\ a, J'b \end{Bmatrix}' - J'\begin{Bmatrix} \bar{A} \\ \bar{a}, b \end{Bmatrix}'$$

and therefore

$$\nabla = J'\bar{\nabla} + \begin{Bmatrix} A \\ u, \nabla' \end{Bmatrix}' - J'\begin{Bmatrix} \bar{A} \\ \bar{u}, \bar{\nabla}' \end{Bmatrix}'$$

whence

$$(88) \qquad D \equiv \nabla - \begin{Bmatrix} A \\ u, \nabla' \end{Bmatrix}' = J'\left(\bar{\nabla} - \begin{Bmatrix} \bar{A} \\ \bar{u}, \bar{\nabla}' \end{Bmatrix}\right) = J'\bar{D};$$

D is therefore a covariant differential operator.

Let v be any contravariant vector and set $d_u v = Su\nabla_\alpha \cdot v_\alpha$; then, if f is any function of x and u,

$$(89) \qquad d_v f \equiv (Sv\nabla_\alpha + Sd_u v\nabla'_\alpha)f_\alpha = (SvD_\alpha + S\delta_u v\nabla'_\alpha)f_\alpha$$

where $\delta_u v$ is the contravariant vector defined by

$$(90) \qquad \delta_u v = d_u v + \begin{Bmatrix} A \\ u,\, v \end{Bmatrix} = Vu.$$

The tensor corresponding to the matrix V is known as the covariant derivative of v.

8.14 Functions of a variable matrix. The general theory of analytic functions of a variable matrix $x = \| \xi_{ij} \|$ is co-extensive with that of n^2 scalar variables and hence is so general as to be void of properties peculiar to matrices. This follows immediately from the obvious relation

$$\xi_{ij} = \sum_{p=1}^{n} e_{pi} x e_{jp}$$

which expresses the (i, j) coordinate as a linear function of x.

The differential operator[6] corresponding to x is

$$(91) \qquad \Delta = \left\| \frac{\partial}{\partial \xi_{ij}} \right\|.$$

It is often convenient to have a special notation for the transverse Δ' and when this is so we shall set

$$(92) \qquad \partial = \left\| \frac{\partial}{\partial \xi_{ji}} \right\| = \Delta'.$$

These operators may stand after their operands and the same convention as was used for subscripts attached to ∇ will also be used for Δ and ∂ when necessary.

The fundamental property of Δ is

$$(93) \qquad df = \operatorname{tr}(dx\Delta')f = \operatorname{tr}(dx\partial)f$$

where f is any function of x and $\operatorname{tr}(A)$ stands for the trace of the matrix A. This result follows immediately from

$$\operatorname{tr}(dx\Delta') \equiv \sum_i \sum_k d\xi_{ik} \frac{\partial}{\partial \xi_{ik}}.$$

[6] This operator first occurs in a paper by Taber (1890, (84)) who however did not make any systematic use of it. Macaulay in a tract published in 1893 (110) but written in 1887 used Δ consistently in applying quaternions to physical problems; he used the notation $\mathfrak{C}$ for Δ. Later Born (385) used the same operator to great effect in his theory of quantum matrices. Turnbull (436) uses Ω for Δ'.

8.15 Differentiation formulae. We collect here the principal formulae of differentiation; in each case the operand is x or x' and the dummy subscript is omitted except when the meaning is ambiguous without it. To simplify the expressions we set α and β for the traces of the matrices a and b, and ξ_r for the trace of x^r.

$$\text{(94)}\qquad \begin{aligned} \Delta axb &= a'b = a'xb'\Delta, &\qquad \Delta ax'b &= \alpha b = bx'a\Delta \\ \Delta' axb &= \alpha b = bxa\Delta', &\qquad \Delta' ax'b &= a'b = a'x'b'\Delta' \end{aligned}$$

$$\text{(95)}\qquad \begin{aligned} \Delta x^r &= x^{r-1} + x'x^{r-2} + (x')^2x^{r-3} + \cdots + (x')^{r-1} = \Delta'(x')^r \\ x^r\Delta &= x^{r-1} + x^{r-2}x' + x^{r-3}(x')^2 + \cdots + (x')^{r-1} = (x')^r\Delta' \\ \Delta' x^r &= nx^{r-1} + \xi_1 x^{r-2} + \xi_2 x^{r-3} + \cdots + \xi_r = x^r\Delta' \end{aligned}$$

$$\text{(96)}\qquad \begin{aligned} \Delta\,\mathrm{tr}\,(axb) &= \Delta\,\mathrm{tr}\,(bax) = a'b' \\ \partial\mathrm{tr}(axb) &= ba \end{aligned}$$

$$\text{(97)}\qquad \Delta\,\mathrm{tr}\,(x^r) \equiv \Delta\xi_r = r(x')^{r-1}, \qquad \partial\,\mathrm{tr}\,(x^r) = rx^{r-1}.$$

$$\text{(98)}\qquad \begin{aligned} \mathrm{tr}\,(\Delta)axb &= \alpha b = \mathrm{tr}\,(\partial)axb \\ \mathrm{tr}\,(\Delta)x^r &= rx^{r-1} = \mathrm{tr}\,(\partial)x^r \end{aligned}$$

$$\text{(99)}\qquad \begin{aligned} \partial\,\mathrm{tr}\begin{Bmatrix} x & a \\ r & s \end{Bmatrix} &= (r+s)\begin{Bmatrix} x & a \\ r-1 & s \end{Bmatrix} = \mathrm{tr}\,(\partial)\begin{Bmatrix} x & a \\ r & s \end{Bmatrix} \\ \Delta\,\mathrm{tr}\begin{Bmatrix} x & a \\ r & s \end{Bmatrix} &= (r+s)\begin{Bmatrix} x' & a' \\ r-1 & s \end{Bmatrix}. \end{aligned}$$

$$\text{(100)}\qquad \begin{aligned} \Delta'_\alpha x_\alpha^{r+1} - x\Delta'_\alpha x_\alpha^r &= \mathrm{tr}\,(x^r) \equiv \xi_r \\ (\Delta_\alpha x_\alpha - x'\Delta_\alpha)\, x_\alpha^r &= (x')^r. \end{aligned}$$

The proofs of these formulae are all very similar and we shall consider here only the most important leaving the remainder to the reader. If $a = \| a_{ij} \|$, $b = \| b_{ij} \|$, then

$$\Delta axb = \left\| \frac{\partial}{\partial \xi_{ip}} a_{pq}\xi_{qr}b_{rj} \right\| = \left\| \sum_p a_{pi}b_{pj} \right\| = a'b;$$

hence also

$$a'x'b'\Delta' = (\Delta bxa)' = (b'a)' = a'b.$$

The remaining parts of (94) follow in the same way. It follows also from (94) that

$$\begin{aligned} \Delta_\alpha x_\alpha^r &= \Delta_\alpha x_\alpha x^{r-1} + \Delta_\alpha x x_\alpha^{r-1} = x^{r-1} + \Delta_\alpha x x_\alpha x^{r-2} + \Delta_\alpha x^2 x_\alpha^{r-2} \\ &= x^{r-1} + x'x^{r-2} + \Delta_\alpha x^2 x_\alpha x^{r-3} + \Delta_\alpha x^3 x_\alpha^{r-3}, \end{aligned}$$

and so on; the remaining parts of (95) follow from (94) in the same way.

To prove (96) we notice first that tr (ab) = tr (ba) and hence in tr $(a_1 a_2 \cdots a_r)$ the factors may be permuted cyclically. Then, if $c = \| c_{ij} \|$,

$$\partial \operatorname{tr}(cx) = \sum_{i,j} e_{ij} \frac{\partial}{\partial \xi_{ji}} \sum_{p,q} c_{pq} \xi_{qp} = \sum c_{ij} e_{ij} = c.$$

Formula (97) follows by repeated application of (96); thus

$$\begin{aligned}\Delta \operatorname{tr}(x^r) &= \Delta_\alpha \operatorname{tr}(x_\alpha x^{r-1}) + \Delta_\alpha \operatorname{tr}(x_\alpha^{r-1} x) \\ &= (x')^{r-1} + \Delta_\alpha \operatorname{tr}(x_\alpha x^{r-1}) + \Delta_\alpha \operatorname{tr}(x_\alpha^{r-2} x^2) \\ &= (x')^{r-1} + (x')^{r-1} + \cdots \\ &= r(x')^{r-1}\end{aligned}$$

The remaining formulae are proved in the same way.

If $\Sigma\alpha_r\lambda^r$ is a scalar power series and $f(x) = \Sigma\alpha_r x^r$, then from (97) and (98)

$$\partial \operatorname{tr}(f(x)) = \Sigma r\alpha_r x^{r-1} = f'(x) = \operatorname{tr}(\partial) f(x)$$

so that the operators ∂ tr ( ) and tr (∂) have the same effect on such functions. Similarly, if $F(x) = \Sigma\alpha_r \begin{Bmatrix} x & a \\ r & s_r \end{Bmatrix}$, it follows from (99) that

(101) $$\partial \operatorname{tr}(F) = \operatorname{tr}(\partial) F = \Sigma(r + s_r) \begin{Bmatrix} x & a \\ r-1 & s_r \end{Bmatrix}.$$

8.16 As an illustration of the application of the formulae of the preceding section we shall give some parts of the theory of quantum matrices which are applicable to matrices of finite order.

Let $q_1, q_2, \cdots, q_f; p_1, \cdots, p_f$ be the coordinates of a dynamical system and $\mathfrak{H}$ the Hamiltonian function; these coordinates satisfy the system of ordinary partial differential equations

(102) $$\dot{q}_i = \frac{\partial \mathfrak{H}}{\partial p_i}, \qquad \dot{p}_i = -\frac{\partial \mathfrak{H}}{\partial q_i} \qquad (i = 1, 2, \cdots, f).$$

We may suppose that $f = n^2$, a perfect square; for, if $(n-1)^2 < f < n^2$, we can introduce $2(n^2 - f)$ additional coordinates $q_{f+1}, p_{f+1}, \cdots, q_{n^2}, p_{n^2}$ which do not occur in $\mathfrak{H}$ so that these variables equated to constants are solutions of the extended system. When this is done, we can order the q's and p's in square arrays $\| q_{ij} \|$, $\| p_{ij} \|$ in such a way that p_{ij} corresponds to q_{ji} for all i and j. Equation (102) then becomes

$$\dot{q}_{ij} = \frac{\partial \mathfrak{H}}{\partial p_{ji}}, \qquad \dot{p}_{ij} = -\frac{\partial \mathfrak{H}}{\partial q_{ji}}$$

or, if the matrices $\| q_{ij} \|$ and $\| p_{ji} \|$ are denoted by q and p and the corresponding transverse differential operators by ∂_q and ∂_p,

(103) $$\dot{q} = \partial_p \mathfrak{H}, \qquad \dot{p} = \partial_q \mathfrak{H}.$$

If we transform (103) by the substitution

$$q = \epsilon^{wt}Q\epsilon^{-wt}, \qquad Q = \| Q_{ij} \|, \qquad p = \epsilon^{wt}P\epsilon^{-wt}, \qquad P = \| P_{ij} \|,$$

where w is a constant matrix, we get

$$\dot{q} = \epsilon^{wt}(\dot{Q} + wQ - Qw)\epsilon^{-wt}, \qquad \dot{p} = \epsilon^{wt}(\dot{P} + wP - Pw)\epsilon^{-wt}.$$

Also, if t is kept constant,

$$\text{tr}\,(dq\partial_q) = \text{tr}\,(\epsilon^{wt}dQ\epsilon^{-wt}\partial_q) = \text{tr}\,(dQ\epsilon^{-wt}\partial_q\epsilon^{wt})$$

with a similar relation for p. Hence

$$\partial_q = \epsilon^{wt}\partial_Q\epsilon^{-wt}, \qquad \partial_p = \epsilon^{wt}\partial_P\epsilon^{-wt}.$$

Using these results in (103) we get

(104) $$(\dot{Q} + wQ - Qw) = \partial_P\mathfrak{H}, \qquad (\dot{P} + wP - Pw) = -\partial_Q\mathfrak{H},$$

$\mathfrak{H}$ being expressed in terms of P and Q and, if necessary, also t. Now from (96)

$$Qw - wQ = \partial_P\text{tr}\,[w(PQ - QP)], \qquad -(Pw - wP) = \partial_Q\text{tr}\,[w(PQ - QP)]$$

and hence, if

(105) $$\mathfrak{K} \equiv \mathfrak{H} + \text{tr}[w(PQ - QP)] = \mathfrak{H} + \text{tr}[w(pq - qp)],$$

we have in place of (103)

(106) $$\dot{Q} = \partial_P\mathfrak{K}, \qquad \dot{P} = -\partial_Q\mathfrak{K},$$

so that the transformation is canonical.

If $\dot{Q} = 0 = \dot{P}$ in (104), then

$$wQ - Qw = \partial_P\mathfrak{H}, \qquad wP - Pw = -\partial_Q\mathfrak{H}$$

or on restoring the exponential factor

(107) $$wq - qw = \partial_p\mathfrak{H}, \qquad wp - pw = -\partial_q\mathfrak{H}.$$

When $\mathfrak{H}$ is given, these are algebraic equations which can be solved for p and q; the solution will of course generally contain arbitrary parameters.

Under the same assumptions (106) becomes

(108) $$\partial_P\mathfrak{K} = 0 = \partial_Q\mathfrak{K},$$

and if P, Q are independent variables, the only solution is $\mathfrak{K}$ = constant and the only solution for $\mathfrak{H}$ in (107) then has the form

$$\mathfrak{H}^* = \text{tr}[w(pq - qp)]$$

apart from an additive constant. Equation (108) may then be written

(109) $$\partial_p(\mathfrak{H} - \mathfrak{H}^*) = 0 = \partial_q(\mathfrak{H} - \mathfrak{H}^*).$$

Now, if $r_1, r_2, \cdots, r_m$ are the parameters in the solution of (107) we have

$$\frac{\partial(\mathfrak{H} - \mathfrak{H}^*)}{\partial r_k} = \sum \frac{\partial p_{ij}}{\partial r_k} \frac{\partial(\mathfrak{H} - \mathfrak{H}^*)}{\partial p_{ij}} + \sum \frac{\partial q_{ij}}{\partial r_k} \frac{\partial(\mathfrak{H} - \mathfrak{H}^*)}{\partial q_{ij}}$$

which vanishes in virtue of (109). Hence, if $\mathfrak{H}$ is expressed in terms of $r_1, r_2, \cdots, r_m$ by using the solutions of (107), it will differ by an additive constant at most from $-\operatorname{tr}[w(pq - qp)]$.

CHAPTER IX

THE AUTOMORPHIC TRANSFORMATION OF A BILINEAR FORM

9.01 If the variables of a bilinear form whose matrix is a are transformed cogrediently by a matrix x, the matrix of the new bilinear form is $x'ax$; when this new form is identical with the old, the transformation is said to be *automorphic*. The problem of finding all automorphic transformations of a is therefore equivalent to solving the equation

$$x'ax = a. \tag{1}$$

We shall assume for the present that $|\,a\,| \neq 0$ in which case also $|\,x\,| \neq 0$.

It follows from (1) that $x'a = ax^{-1}$. Hence, if $f(\lambda)$ is a scalar polynomial

$$[f(x)]'a = f(x')a = af(x^{-1}). \tag{1'}$$

In particular, if $f(\lambda) = (1 - \lambda)/(1 + \lambda)$ and $y = f(x^{-1})$, then

$$y = \frac{1 - x^{-1}}{1 + x^{-1}} = \frac{x - 1}{x + 1} = -f(x) \tag{2}$$

provided $|\,x + 1\,| \neq 0$. Hence from (1′) $y'a = -ay$ so that

$$y' = -aya^{-1}. \tag{3}$$

Conversely, if y satisfies (3) and $|\,1 - y\,| \neq 0$, then $x = (1 + y)/(1 - y)$ is a solution of (1) such that $|\,x + 1\,| \neq 0$. For from (3) $f(y')a = af(-y)$ so that

$$x'a = \frac{1 + y'}{1 - y'}a = a\frac{1 - y}{1 + y} = ax^{-1}.$$

Similarly, if $|\,x - 1\,| \neq 0$, we may set $x = (1 - y)/(1 + y)$ and then y is a solution of (3) such that $|\,1 + y\,| \neq 0$, and conversely. The effect of the transformation (2) is therefore to reduce the solution of (1), which is quadratic in x, to that of (3), which is linear in y, except when both 1 and -1 are roots of x. It is because (3) is linear that it is more convenient than (1); in particular if we regard

$$y'a + ay = 0, \qquad y = \|\,\eta_{pq}\,\|, \qquad a = \|\,\alpha_{pq}\,\|, \tag{3'}$$

as a system of n^2 linear homogeneous equations in the η's, then the rank of the system gives the number of parameters which enter into the solution when those values of y (or of x) are excluded for which both 1 and -1 are roots.

Since the main problem is thus reduced to the solution of linear equations, it may be regarded as solved; the solution, however, can be given a somewhat more definite form as we shall now show.

9.02 The equation $y' = \pm\, aya^{-1}$. We shall consider in place of (3) the more general equation

(4) $$y' = \delta aya^{-1}, \qquad \delta = \pm 1.$$

Forming the transverse we get $y = \delta a'^{-1}y'a'$ or $y' = \delta a'ya'^{-1}$, whence

(5) $$ya^{-1}a' = a^{-1}a'y, \qquad a'ya'^{-1} = aya^{-1}$$

so that y is commutative with $a^{-1}a'$. Now from (4) we have $y = \delta a^{-1}y'a$ and hence

$$2y = y + \delta a^{-1}y'a.$$

But if b is any matrix commutative with $a^{-1}a'$, then

(6) $$y = b + \delta a^{-1}b'a$$

is a solution of (4); for on substituting this value of y we get

$$y' - \delta aya^{-1} = b' + \delta a'ba'^{-1} - \delta aba^{-1} - b' = 0$$

since, as in (5), $a'ba'^{-1} = aba^{-1}$. It was noted above that y has this form and it therefore follows that the general solution of (3) is obtained by setting

(7) $$y = b - a^{-1}b'a, \qquad ba^{-1}a' = a^{-1}a'b.$$

It should be noted, however, that two different values of b may give rise to the same value of y.

9.03 We are now able to give a solution of (1) under the restriction that either $|\, x + 1\, | \neq 0$ or $|\, x - 1\, | \neq 0$. Since the first condition is transformed into the second if x is changed into $-x$, it is sufficient for the present to assume that $|\, x + 1\, | \neq 0$, and in this case the value of y given by (2) is finite. In terms of y we have

$$x = \frac{1+y}{1-y} = \frac{1 + b - a^{-1}b'a}{1 - b + a^{-1}b'a} = (a - ab + b'a)^{-1}(a + ab - b'a)$$

or, if

(8) $$c = ab - b'a,$$

then

(9) $$\begin{aligned} x &= (a - c)^{-1}(a + c), &\qquad |\, x + 1\, | \neq 0, \\ x &= (c - a)^{-1}(a + c), &\qquad |\, x - 1\, | \neq 0. \end{aligned}$$

It follows as in §9.01 that, if x has this form, it is a solution of (1).

In place of (8) we may define c by

(10) $$c' = -a'a^{-1}c = -ca^{-1}a'.$$

For from (7) and (8)

$$c' = b'a' - a'b = -a'a^{-1}c = -ca^{-1}a'$$

and, if c is given by (10) and x by (9), then

$$\begin{aligned} x' &= (a' + c')\,(a' - c')^{-1} = (a' - ca^{-1}a')\,(a' + ca^{-1}a')^{-1} \\ &= (1 - ca^{-1})\,(1 + ca^{-1})^{-1} = (1 + ca^{-1})^{-1}\,(1 - ca^{-1}) = a(a + c)^{-1}\,(a - c)a^{-1} \\ &= ax^{-1}a^{-1}. \end{aligned}$$

If a is symmetric, (8) or (10) gives $c' = -c$, and c is otherwise arbitrary except that $|\,a - c\,| \neq 0$; in particular if $a = 1$, (9) reduces to the form of an orthogonal matrix already given in 6.03. Similarly if a is skew, (10) shows that c is an arbitrary symmetric matrix subject to the condition that $|\,a - c\,| \neq 0$.

The case in which a is symmetric can also be handled as follows. We can set $a = b^2$ where b is symmetric and, if

$$y = b^{-1}xb,$$

equation (1) gives $yy' = 1$ so that y is orthogonal. Conversely, if y is any orthogonal matrix $x = byb^{-1}$ is a solution of (1).

9.04 **Principal idempotent and nilpotent elements.** Since x is similar to $(x')^{-1}$, the elementary divisors which correspond to roots other than ± 1 occur in pairs with reciprocal roots. If we arrange these roots in pairs g_r, g_r^{-1} and denote the corresponding principal idempotent elements by e_r and e_{-r}, respectively, we may set

$$(11)\qquad x = \Sigma\,[g_r(e_r + \xi_r) + g_r^{-1}\,e_{-r}(1 + \xi_{-r})^{-1}] + \theta_1(e_1 + \xi_1) - \theta_2(e_{-1} + \xi_{-1})$$

where the ξ's are nilpotent, e_1, e_{-1} are the principal idempotent elements corresponding to 1 and -1, if present as roots, and θ_1, θ_2 are either 0 or 1. The form of x^{-1} is then

$$(12)\quad x^{-1} = \Sigma\,[g_r(e_{-r} + \xi_{-r}) + g_r^{-1}\,e_r(1 + \xi_r)^{-1}] + \theta_1 e_1(1 + \xi_1)^{-1} - \theta_2 e_{-1}(1 + \xi_{-1})^{-1}$$

and (11) gives

$$(13)\qquad \begin{aligned} e_r' &= ae_{-r}a^{-1}, & e_{-r}' &= ae_ra^{-1} \\ \xi_r' &= a\xi_{-r}a^{-1}, & \xi_{-r}' &= a\xi_ra^{-1} \end{aligned}$$

$$(13')\qquad \begin{aligned} e_1' &= ae_1a^{-1}, & e_{-1}' &= ae_{-1}a^{-1} \\ \xi_1' &= -a\,\frac{\xi_1}{1 + \xi_1}\,a^{-1}, & \xi_{-1}' &= -a\,\frac{\xi_{-1}}{1 + \xi_{-1}}\,a^{-1}. \end{aligned}$$

We require also the form of $x + x^{-1}$ and $x - x^{-1}$; if

$$(14)\qquad \alpha_r = \xi_r + \xi_{-r}, \qquad \beta_r = \xi_r - \xi_{-r}$$

then

$$\begin{aligned} x + x^{-1} &= \Sigma\,[g_r(e_r + e_{-r} + \alpha_r) + g_r^{-1}\,(e_r + e_{-r})\,(1 + \alpha_r)^{-1}] \\ &\quad + \theta_1[(e_1 + \xi_1) + e_1(1 + \xi_1)^{-1}] - \theta_2[(e_{-1} + \xi_{-1}) + e_{-1}(1 + \xi_{-1})^{-1}] \end{aligned}$$

$$\begin{aligned} x - x^{-1} &= \Sigma\,[g_r(e_r - e_{-r} + \beta_r) - g_r\,(e_r - e_{-r})\,(1 + \beta_r)^{-1}] \\ &\quad + \theta_1[(e_1 + \xi_1) - e_1(1 + \xi_1)^{-1}] - \theta_2[(e_{-1} + \xi_{-1}) - e_{-1}(1 + \xi_{-1})^{-1}] \end{aligned}$$

or if

$$\gamma_r = g_r\alpha_r - g_r^{-1}\frac{\alpha_r}{1+\alpha_r}, \qquad \gamma_1 = \frac{\xi_1^2}{1+\xi_1}, \qquad \gamma_{-1} = \frac{\xi_{-1}^2}{1+\xi_{-1}}$$

$$\delta_r = g_r\xi_r + g_r^{-1}\frac{\xi_r}{1+\xi_r}, \qquad \delta_{-r} = g_r\xi_{-r} + g_r^{-1}\frac{\xi_{-r}}{1+\xi_{-r}}, \tag{15}$$

$$\delta_1 = 2\xi_1 - \gamma_1, \qquad \delta_{-1} = 2\xi_{-1} - \gamma_{-1}$$

we have

$$x + x^{-1} = \Sigma\,[(g_r + g_r^{-1})\,(e_r + e_{-r}) + \gamma_r] + \theta_1(2e_1 + \gamma_1) - \theta_2(2e_{-1} + \gamma_{-1}) \tag{16}$$

$$x - x^{-1} = \Sigma\,[\{(g_r - g_r^{-1})\,e_r + \delta_r\} - \{(g_r - g_r^{-1})\,e_{-r} + \delta_{-r}\}] + \theta_1\delta_1 - \theta_2\delta_{-1} \tag{17}$$

where the elements grouped together are principal elements.

The principal idempotent elements of $x - x^{-1}$ are also principal idempotent elements of x except that roots 1 and -1 of x both give the same root 0 of $x - x^{-1}$; no root of x other than ± 1 leads to the coalescing of roots in $x - x^{-1}$. If we put

$$2u = x + x^{-1}, \qquad 2v = x - x^{-1}, \tag{18}$$

then u is a solution of (4) with $\delta = 1$ and v is a solution with $\delta = -1$; also

$$x^2 - 2vx - 1 = 0 \tag{19}$$

which has the formal solution

$$x = v + (v^2 + 1)^{\frac{1}{2}}. \tag{20}$$

Here $v^2 + 1 = u^2$ so that $(v^2 + 1)^{\frac{1}{2}}$ exists whenever x is a solution of (1). Conversely, if v_1 is any solution of (4) with $\delta = -1$ and if u_1 is a determination of $(v_1^2 + 1)^{\frac{1}{2}}$ such that

$$u_1'a = au_1, \tag{21}$$

then x is a solution of (1); for

$$x'a = v_1'a + u_1'a = -av_1 + au_1 = ax^{-1}$$

since

$$(u_1 + v_1)\,(u_1 - v_1) = u_1^2 - v_1^2 = 1.$$

If $v^2 + 1$ has no zero root, determinations of $(v^2 + 1)^{\frac{1}{2}}$ always exist which are polynomials in v^2 and therefore satisfy (21); but even in this case this does not give all solutions. The situation is as follows. The general form of v is given by (17) if we replace $g_r - g_r^{-1}$ by, say, $2k_r$ and $\theta_1\delta_1 - \theta_2\delta_{-1}$ by δ_0. When k_r is given, then g_r is determined; from (13) and (14) we have $(\delta_r' + \delta_{-r}')a = a(\delta_r + \delta_{-r})$ and therefore, if $k_r^2 + 1 \neq 0$, the part of $(v^2 + 1)^{\frac{1}{2}}$ corresponding to $e_r + e_{-r}$ exists and satisfies (21); we therefore get all valid expressions for this part of $(v^2 + 1)^{\frac{1}{2}}$ by using the form of the square root given in §8.05 with the restriction that the only sets of partial units that may be used are those that satisfy (21). However, since $(v^2 + 1)^{\frac{1}{2}} = u$, (16) and (17) show that we need only use the idempotent elements $e_r + e_{-r}$, which are determined by v, in those parts of the square root which do not depend on the zero root of v; e_1 and e_{-1}, however, are not defined by v so that it is necessary in any particular case to consider how e_0 and δ_0 can be broken up into parts which have the required property.

If v has a zero root with the principal idempotent and nilpotent parts e_0, δ_0, then $e^2 = e$ shows that, although $\delta_0' a = -a\delta_0$, we have

$$(21') \qquad e_0' a = ae_0.$$

We therefore seek to divide e_0 into two idempotent parts, e_1 and e_{-1}, which are commutative with v and therefore with δ_0. In forming the square root we then attach the value $+1$ to e_1 and -1 to e_{-1}.

If $k_r^2 + 1 = 0$, then $g_r = i$ and the corresponding part of $(v^2 + 1)^{\frac{1}{2}}$ is $2i(\delta_r + \delta_{-r}) + \delta_r^2 + \delta_{-r}^2$, and it is readily shown from (15) that this has a square root. The details are left to the reader.

If b is a solution of $b'a = -ab$, then so are also $t = \tan b$ and $v = \tan 2b$. A short calculation then gives $x = (1 + t)/(1 - t)$ subject to the restrictions already given; this shows the relations between the rational and irrational solutions.

9.05 **The exponential solution.** Some of the difficulties of the solution in §9.04 can be avoided by setting

$$(22) \qquad x = \exp(z) \equiv \epsilon^z, \qquad z = \operatorname{Log} x$$

where a principal determination of $\log x$ is to be used. Since this determination of $\log x$ is a polynomial in x and $x'ax = a$, we have

$$(23) \qquad z' = \operatorname{Log} x' = \operatorname{Log} ax^{-1}a^{-1} = a(\operatorname{Log} x^{-1})a^{-1}$$

and therefore

$$(24) \qquad 1 = x'axa = \epsilon^{z'}\epsilon^{aza^{-1}} = \epsilon^{z' + aza^{-1}}.$$

From (11)

$$(25) \qquad \begin{aligned} z &= \Sigma\,[(\log g_r)\, e_r + \eta_r - (\log g_r)\, e_{-r} + \eta_{-r}] + \theta_1\eta_1 + \theta_2(\pi i e_{-1} + \eta_{-1}) \\ \eta_s &= \xi_s - \tfrac{1}{2}\xi_s^2 + \cdots \qquad (s = r, -r, 1, -1) \end{aligned}$$

and from (13)

$$(26) \qquad \begin{aligned} e_s' &= ae_{-s}a^{-1}, \qquad \eta_s' = -a\eta_{-s}a^{-1} \qquad (s = r, -r), \\ e_s' &= ae_s a^{-1}, \qquad \eta_s' = -a\eta_s a^{-1} \qquad (s = 1, -1). \end{aligned}$$

Hence

$$\begin{aligned} z' + aza^{-1} &= \Sigma\,[(\log g_r)\, e_r' + \eta_r' - (\log g_r)\, e_{-r}' + \eta_{-r}'] + \theta_1\eta_1' + \theta_2(\pi i e_{-1}' + \eta_{-1}') \\ &\quad + \Sigma\,[(\log g_r)\, e_{-r}' - \eta_{-r}' - (\log g_r) e_r' - \eta_r'] - \theta_1\eta_1' + \theta_2(\pi i e_{-1}' - \eta_{-1}') \\ &= 2\theta_2\pi i e_{-1}', \end{aligned}$$

and therefore, if we set

(27) $$\zeta = \theta e_{-1}\,(\theta \equiv \theta_2 = 0, 1), \qquad w = z - \pi i \zeta,$$

we have

(28) $$w' + awa^{-1} = 0, \qquad \zeta' - a\zeta a^{-1} = 0,$$

and

(29) $$w = \Sigma\,[(\log g_r)e_r + \eta_r - (\log g_r)e_{-r} + \eta_{-r}] + \theta_1\eta_1 + \theta_2\eta_{-1}.$$

The general value of x can therefore be expressed in terms of the solution of the equation discussed in §9.02.

If we now start with w as a solution of (28) and define x by $x = \epsilon^w$, then

$$x' = \epsilon^{w'} = \epsilon^{-awa^{-1}} = a\epsilon^{-w}a^{-1} = ax^{-1}a^{-1}$$

and therefore x is a solution of (1); to obtain every solution, however, we must add the terms $\pi i \zeta$ to w.

If e_0 is the principal idempotent element corresponding to the root 0, then (29) shows that the presence of the ζ-term depends on the division of e_0 into two parts e_1 and e_{-1} which satisfy the second set of equations in (26); and corresponding to these we have nilpotent parts η_1 and η_{-1} which give rise to 1 and -1, respectively, as roots of x, or 0, πi as roots of z.

A form which gives rational parameters is obtained from the exponential solution as follows. Let

(30) $$t = \tanh(z/2) = \frac{\epsilon^z - 1}{\epsilon^z + 1} = \frac{x - 1}{x + 1}$$

then

(31) $$x = \frac{1 + t}{1 - t},$$

and

$$ata^{-1} = \frac{axa^{-1} - 1}{axa^{-1} + 1} = \frac{1 - ax^{-1}a^{-1}}{1 + ax^{-1}a^{-1}} = \frac{1 - x'}{1 + x'} = -t'$$

so that (31) gives a solution of (1). If, however, $|\,x + 1\,| = 0$, then t becomes infinite so that (31) cannot give directly any x which has -1 as a root. This difficulty arises from the fact that $\tanh(\theta/2) \to \infty$ when $\theta \to \pi i$; but, since $(t + 1)/(t - 1) = \epsilon^z$ for all values of t which do not have an infinite root, that is, one corresponding to a root $(2k + 1)\pi i$ of z, hence x will be a solution of (1) so long as the coordinates of z are continuous functions of the parameters involved and the limiting value of x is finite and determinate.

9.06 **Matrices which admit a given transformation.** In (1) we may regard x as given and a as unknown; the problem then is to find all matrices a such that

(32) $$x'ax = a.$$

If we associate with $a = \| a_{ij} \|$ the corresponding tensor of grade 2,

$$u = \Sigma a_{ij} e_i e_j,$$

we see immediately that (23) corresponds to setting (cf. §5.10)

$$\Pi_2(x)u = u. \tag{33}$$

Hence there is a solution if, and only if, x has at least one pair of reciprocal roots; in this case $\Pi_2(x)$ has one or more roots equal to 1 and the various invariant elements corresponding to this root give a linearly independent set of determinations of a.

When it is required that $| a | \neq 0$, another form of solution is preferable. In this case $x' = ax^{-1}a^{-1}$; but, since x and x' are similar, we also have $x' = pxp^{-1}$, where, if p_1 is one determination of p, the general form is

$$p = p_1 b, \qquad bx = xb, \qquad | b | \neq 0. \tag{34}$$

Hence it is necessary that x^{-1} be similar to x, say

$$x^{-1} = q_1^{-1} x q_1, \tag{35}$$

which gives immediately

$$a = p_1 b q_1. \tag{36}$$

Conversely, if p_1, b, and q_1 satisfy the given conditions, it follows immediately that (36) gives a solution of (32).

CHAPTER X

LINEAR ASSOCIATIVE ALGEBRAS

10.01 **Fields and algebras.** A set of elements which are subject to the laws of ordinary rational algebra is called a *field*. We may make this idea more precise as follows. Let $a, b, \cdots$ be a set of entities, F, which are subject to two operations, addition and multiplication; this set is called a field if it satisfies the following postulates:[1]

A1. $a + b$ is a uniquely determined element of F.

A2. $a + b = b + a$.

A3. $(a + b) + c = a + (b + c)$.

A4. There is a unique element 0 in F such that $a + 0 = a$ for every element a in F.

A5. For every element a in F there exists a unique element b in F such that $a + b = 0$.

M1. ab is a unique element of F.

M2. $ab = ba$.

M3. $ab\cdot c = a\cdot bc$.

M4. There is a unique element 1 in F such that $a1 = a$ for every a in F.

M5. For every element $a \neq 0$ in F there exists a unique element b in F such that $ab = 1$.

AM. $a(b + c) = ab + ac$, $(b + c)a = ba + ca$.

R. If m is a whole number and ma denotes the element which results from adding together m a's, then $ma \neq 0$ for any $m > 0$ provided that $a \neq 0$.

If M2 is omitted the resulting set is said to be a *division algebra*. This does not imply that M2 does not hold, only that it is not presupposed; if it does hold, the algebra is said to be commutative. If M2, 4, 5 are all omitted, the corresponding set is called an *associative algebra*. If the algebra contains an identity, that is, an element satisfying the condition laid down in M4 for 1, this element is called the *principal unit* of the algebra. Postulate R is included merely as a matter of convenience; its effect is to exclude modular fields. In consequence of R every field which we shall consider contains[2] the field of rational numbers as a subset.

As an example of a field we may take the field of rational numbers extended by a cube root of unity, $\omega = (-1 + \sqrt{-3})/2$. Every number of this field can be put in the form

$$a = \alpha + \beta\omega = \alpha 1 + \beta\omega$$

[1] These postulates are not independent; they are formed so as to show the principal properties of the set. In place of M5 it is often convenient to take: M5′ If $a \neq 0$, $ax = 0$ implies $x = 0$.

[2] Strictly speaking, we should say that the field contains a subset simply isomorphic with the field R of rational numbers. This subset is then used in place of R in the same way as scalars are replaced by scalar matrices in §1.04.

where α and β are rational numbers; the form of a is unique since $\alpha + \beta\omega = \gamma + \delta\omega$ gives $(\beta - \delta)\,\omega = \gamma - \alpha$ and, since ω is not rational, this is impossible unless $\beta - \delta = 0 = \gamma - \alpha$. We say that $1, \omega$ is a *basis* of F relative to the field R of rational numbers, and F is said to be a field of order 2 over R.

As an example of an associative algebra we may take the algebra of matrices with rational coordinates. Here any element a of the algebra can be put uniquely in the form $a = \Sigma a_{ij}e_{ij}$, where the a_{ij} are rational numbers; and $e_{ij}(i, j = 1, 2, \cdots, n)$ form a basis of the algebra, which is of order n^2. We also have an algebra if the coordinates a_{ij} are taken to be any elements of the field $F = (1, \omega)$ described above. This algebra is one of order n^2 over F. Instead of regarding it as an algebra over F we may clearly look on it as an algebra of order $2n^2$ over R the basis being $e_{ij}, \omega e_{ij}(i, j = 1, 2, \cdots, n)$.

10.02 Algebras which have a finite basis. Let A be a set of elements which form an associative algebra and G a subset which is also an algebra. We shall say that $a_1, a_2, \cdots, a_n$ form a basis of A relatively to G if (i) each a_i lies in A, (ii) if every element of A can be put uniquely in the form

$$a = \gamma_1 a_1 + \gamma_2 a_2 + \cdots + \gamma_n a_n \tag{1}$$

where the γ's belong to G. Though it is not altogether necessary to do so, we shall restrict ourselves to the case in which G is a field which contains the rational field, that is, we assume as a postulate:

BR. For every algebra A under consideration there exists a non-modular field F and a subset of elements $a_1, a_2, \cdots, a_n$ such that (i) every element of A can be put uniquely in the form

$$a = \sum_1^n \gamma_i a_i \qquad (\gamma_i \text{ in } F)$$

and (ii) every element of this form belongs to A; and further the elements of F are commutative with $a_1, a_2, \cdots a_n$.

Since the product of any two elements of A is also an element of A and can therefore be expressed in the form (1), we have

$$a_i a_j = \sum_k \gamma_{ijk} a_k \qquad (i, j = 1, 2, \cdots, n) \tag{2}$$

where γ_{ijk} are elements of F. Since the law of combination of the elements of F is supposed known, (2) defines the product of any two elements of A; for

$$(\Sigma\alpha_i a_i)\,(\Sigma\beta_i a_i) = \Sigma\alpha_i\beta_j a_i a_j = \Sigma\alpha_i\beta_j\gamma_{ijk}a_k. \tag{3}$$

If the values of the γ's are assigned arbitrarily in F, it is readily shown that the only postulate which is possibly violated is M3 which states that $ab\cdot c = a\cdot bc$; and in order that this condition shall be satisfied it is necessary and sufficient that $a_i \cdot a_j a_k = a_i a_j \cdot a_k$ for all the elements of the basis. This gives immediately the 'associativity' condition

$$\sum_{\alpha} \gamma_{jk\alpha}\gamma_{i\alpha l} = \sum_{\alpha} \gamma_{ij\alpha}\gamma_{\alpha kl} \qquad (i, j, k, l = 1, 2, \cdots, n). \tag{4}$$

10.03 **The matric representation of an algebra.** If we set

$$A_i = \sum_{p,q=1}^{n} \gamma_{iqp} e_{pq} \qquad (i = 1, 2, \cdots, n), \tag{5}$$

the law of multiplication for matrices gives

$$A_i A_j = \Sigma \gamma_{i\alpha p}\gamma_{jq\alpha} e_{pq}$$

and therefore from (4)

$$A_i A_j = \Sigma \gamma_{ij\alpha}\gamma_{\alpha qp} e_{pq} = \Sigma \gamma_{ij\alpha} A_\alpha.$$

Hence the set of matrices of the form $\Sigma \alpha_i A_i$ is isomorphic with the given algebra in regard to both addition and multiplication. Further, if the algebra contains the identity, the isomorphism is simple; for, if there exist elements α_i of the field such that $\Sigma \alpha_i A_i = 0$, it follows that

$$(\Sigma \alpha_i a_i)x = 0$$

for every element x of the algebra, and putting $x = 1$ we get $\Sigma \alpha_i a_i = 0$.

If the algebra does not have a principal unit, all that is necessary is to replace (5) by

$$A_i = \sum_{p,q=1}^{n+1} \gamma_{iqp} e_{pq} \tag{6}$$

where $\gamma_{i,j,n+1} = 0$ $(i, j \leq n)$ and $\gamma_{n+1,i,j} = \delta_{ij} = \gamma_{i,n+1,j}$ for all i and j.

The importance of this representation is that it enables us to carry over the theory of the characteristic and reduced equations from the theory of matrices. The main theorem is as follows.

THEOREM 1. *The general element $x = \Sigma \xi_i a_i$ satisfies an equation of the form*

$$\lambda^m + b_1\lambda^{m-1} + \cdots + b_m = 0 \tag{7}$$

where b_p is a rational homogeneous polynomial in the ξ's of degree p; and if the variable coordinates ξ_p are given particular values in F, there exists a rational polynomial

$$\varphi(\lambda) = \lambda^\mu + \beta_1\lambda^{\mu-1} + \cdots + \beta_\mu \tag{8}$$

such that (i) $\varphi(x) = 0$, (ii) if $\psi(\lambda)$ is any polynomial with coefficients in F such that $\psi(x) = 0$, then $\varphi(\lambda)$ is a factor of $\psi(\lambda)$.

This theorem follows immediately from the theory of the reduced equation as given in §2.05 and from the fact that the equation which is satisfied by the general element must clearly be homogeneous in the coordinates of that element.

As in the theory of matrices, $-b_1$ is called the *trace* of x and is written $\text{tr}(x)$. The trace is linear and homogeneous in the coordinates and hence $\text{tr}(x + y) = \text{tr}(x) + \text{tr}(y)$.

10.04 **The calculus of complexes.** If $x_1, x_2, \cdots, x_r$ are any elements of an algebra A in a field F, the set B of all elements of the form $\Sigma\xi_i x_i$ (ξ_i in F) is called a *complex*[3] or linear set. Any subset B of A which has the property that, when x, y are any two of its elements, then $\xi x + \eta y$ is also an element of the set is a complex. This follows readily from the theory of linear dependence and the existence of a finite basis for A; it is also easily shown that any subcomplex of A has a finite basis; the order of this basis is called the order of the complex.

We shall write $B = (x_1, x_2, \cdots, x_r)$; this does not imply that the x's are necessarily linearly independent. If $C = (y_1, y_2, \cdots, y_s)$ is a second complex, the *sum* of B and C is defined by

$$B + C = (x_1, x_2, \cdots, x_r, y_1, y_2, \cdots, y_s),$$

that is, $B + C$ is the set of all elements of the form $x + y$ where x lies in B and y in C. Similarly the *product* is defined by

$$BC = (x_i y_j : i = 1, 2, \cdots, r; j = 1, 2, \cdots, s).$$

The set of elements common to B and C forms a complex called the *intersection* of B and C; it is denoted by $B \frown C$. If B and C have no[4] common element, we write $B \frown C = 0$. If every element of C lies in B but not every element of B in C, we shall write $C < B$; in this case $B + C = B$. A complex of order 1 is defined by a single element, say x_1, and for most purposes it is convenient to denote the complex (x_1) simply by x_1; $x_1 < B$ then means that x_1 is an element of B.

If a complex B is an algebra, the product of any two of its elements lies in B and hence $B^2 \leq B$; conversely, if this condition is satisfied, the definition of the product $BB \equiv B^2$ shows that B is an algebra.

We add a summary of the properties of the symbols introduced in this section.

$$B + C = C + B, \quad (B + C) + D = B + (C + D), \quad BC \cdot D = B \cdot CD,$$
$$B \frown C = C \frown B, \quad (B \frown C) \frown D = B \frown (C \frown D),$$
$$B(C + D) = BC + BD, \quad (C + D)\, B = CB + DB,$$
$$B + (C \frown D) \leq (B + C) \frown (B + D), \quad B(C \frown D) \leq BC \frown BD.$$

[3] The term 'complex,' which was introduced by Frobenius in the theory of groups, is more convenient than 'linear set' and no confusion is likely to arise between this meaning of the term and the one used in geometry.

[4] To avoid circumlocution we say the complexes have 'no element in common' in place of the more correct phrase 'no element in common except 0.'

If $B \leq C$, then $B + C = C$, and conversely.

If $B < C$, there exists $D < C$ such that $C = B + D$, $B \frown D = 0$.

If $B = C + D$ and $C \frown D = 0$, we shall say that B is congruent to C modulo D, or

$$B = C \pmod{D};$$

and if b, c, d are elements of B, C, D, respectively, such that $b = c + d$, then

$$b = c \pmod{D}, \qquad c = b \pmod{D}.$$

10.05 **The direct sum and product.** If $A = (a_1, a_2, \cdots, a_\alpha)$ and $B = (b_1, b_2, \cdots, b_\beta)$ are associative algebras of orders α, β, respectively, over the same field F, we can define a new algebra in terms of them as follows. Let C be the set of all pairs of elements (a, b) where $a < A$ and $b < B$ and two pairs (a, b), (a', b') are regarded as equal if, and only if, $a = a'$, $b = b'$. If we define addition and multiplication by

$$(9) \qquad \begin{aligned} (a, b) + (a', b') &= (a + a', b + b') \\ (a, b)\,(a', b') &= (aa', bb') \\ \xi(a, b) &= (\xi a, \xi b) \qquad (\xi \text{ in } F), \end{aligned}$$

it is readily shown that the set C forms an associative algebra. This algebra is called the *direct sum* of A and B and is denoted by $A \oplus B$; its order is $\alpha + \beta$.

The set $\mathfrak{A}$ of all elements of the form $(a, 0)$ forms an algebra which is simply isomorphic with A, and the set $\mathfrak{B}$ of elements $(0, b)$ forms an algebra which is simply isomorphic with B; also

$$C = \mathfrak{A} + \mathfrak{B}, \qquad \mathfrak{A}\mathfrak{B} = 0 = \mathfrak{B}\mathfrak{A}, \qquad \mathfrak{A} \frown \mathfrak{B} = 0.$$

In consequence of this it is generally convenient to say that C is the direct sum of $\mathfrak{A}$ and $\mathfrak{B}$.

If we replace (9) by

$$(9') \qquad \begin{aligned} \xi(a, b) &= (\xi a, b) = (a, \xi b) \qquad (\xi \text{ in } F) \\ (a, b)\,(a', b') &= (aa', bb'), \end{aligned}$$

we get another type of algebra of order $\alpha\beta$ which is called the *direct product* of A and B and is denoted by $A \otimes B$ or by $A \times B$ when there is no chance of confusion. If both A and B contain the identity, the set $\mathfrak{A}$ of elements of the form $(a, 1)$ forms an algebra simply isomorphic with A and the set $\mathfrak{B}$ of elements $(1, b)$ is an algebra simply isomorphic with B; also[5]

$$C = \mathfrak{A}\mathfrak{B} = \mathfrak{B}\mathfrak{A}, \qquad \mathfrak{A} \frown \mathfrak{B} = (1, 1) = 1,$$

and the order of C is the product of the orders of $\mathfrak{A}$ and $\mathfrak{B}$. As in the case of the direct sum it is convenient to say that A is the direct product of $\mathfrak{A}$ and $\mathfrak{B}$ and to indicate this by writing $C = \mathfrak{A} \times \mathfrak{B}$.

[5] Strictly speaking we should use different symbols here for the identity elements of the separate algebras.

The following theorem gives an instance of the direct product which we shall require later.

THEOREM 2. *If an algebra A, which contains the identity, contains also the matric algebra M $(e_{ij}; i, j = 1, 2, \cdots, n)$, the identity being the same for A and M, then A can be expressed as the direct product of M and another algebra B.*

Let B be the set of elements of A which are commutative with every element of M; these elements form an algebra since, if $b_i e_{pq} = e_{pq} b_i$ $(i = 1, 2, \cdots)$, then also

$$(b_i + b_j)e_{pq} = e_{pq}(b_i + b_j), \qquad b_i b_j e_{pq} = e_{pq} b_i b_j.$$

Further $B \frown M$ is the field F, since scalars are the only elements of M which are commutative with every element of M.

If x is any element of A and

$$x_{pq} = \sum_i e_{ip} x e_{qi},$$

then

$$x_{pq} e_{rs} = \sum_i e_{ip} x e_{qi} e_{rs} = e_{rp} x e_{qs} = e_{rs} \sum_i e_{ip} x e_{qi} = e_{rs} x_{pq}$$

so that x_{pq} belongs to B. Also

$$\sum_{pq} x_{pq} e_{pq} = \sum_{pqi} e_{ip} x e_{qi} e_{pq} = \sum_{pq} e_{pp} x e_{qq} = x$$

so that $A = BM$, which proves the theorem.

10.06 Invariant subalgebras. If B is a subalgebra of A such that

$$AB \leq B, \qquad BA \leq B, \tag{10}$$

then B is called an *invariant* subalgebra of A. If we set

$$A = B + C, \qquad B \frown C = 0,$$

the product of any two elements c_i, c_j of C lies in A and hence

$$c_i c_j = c_{ij} + b_{ij}, \qquad c_{ij} < C, \qquad b_{ij} < B.$$

If we now introduce a new operation $\times$ defined by

$$c_i \times c_j = c_{ij}, \tag{11}$$

then the operations $+$ and $\times$, when used to combine elements of C, satisfy all the postulates for an associative algebra. To prove this we need only consider the associativity postulate M3 since the proofs of the others are immediate. If c_1, c_2, c_3 are any elements of C, then both $c_1 \times (c_2 \times c_3)$ and $(c_1 \times c_2) \times c_3$ differ by an element of B from $c_1 c_2 c_3$; their difference is therefore an element of both B and C and hence is 0.

The elements of C therefore form an associative algebra relatively to the operations $+$ and $\times$. When this algebra is considered abstractly, the operation $\times$ may be called multiplication; the resulting algebra is called the *difference* algebra of A and B and is denoted by $(A - B)$.

The difference algebra may also be defined as follows. Let $b_1, b_2, \cdots, b_\beta$ be a basis of B and $c_1, c_2, \cdots, c_\gamma$ a basis of C, so that $b_1, b_2, \cdots, b_\beta, c_1, \cdots, c_\gamma$ is a basis of A. Since A is an algebra, the product c_ic_j can be expressed in terms of this basis and we may therefore set

$$c_ic_j = \Sigma\gamma_{ijk}c_k + \Sigma\delta_{ijk}b_k. \tag{12}$$

The argument used above then shows that

$$d_id_j = \Sigma\gamma_{ijk}d_k \tag{13}$$

defines an associative algebra when B is invariant.

It is readily seen that the form of the difference algebra is independent of the particular complex C which is used to supplement B in A. For if $A = B + P$, $B \frown P = 0$, it follows that to an element p of P there corresponds an element c of C such that $p - c < B$; and we may therefore choose a basis for P for which

$$p_i = c_i + q_i \qquad (q_i < B;\ i = 1, 2, \cdots, \gamma).$$

Equation (12) then gives

$$p_ip_j = \Sigma\gamma_{ijk}p_k + b_{ij}$$

where

$$b_{ij} = q_iq_j + q_ic_j + c_iq_j + \Sigma\delta_{ijk}b_k - \Sigma\gamma_{ijk}b_k < B,$$

and the algebra derived from this in the same way as (13) is from (12) is abstractly the same as before.

If the algebra A does not contain the identity, it may happen that $A^2 < A$, $A^3 < A^2$, and so on. Since the basis of A is finite, we must however have at some stage

$$A^m < A^{m-1}, \qquad A^{m+1} = A^m;$$

the integer m is then called the *index* of A. The most interesting case is when $A^m = 0$; the algebra is then said to be *nilpotent.*

When N_1 and N_2 are nilpotent subalgebras of A which are also invariant, then $N_1 + N_2$ is a nilpotent invariant subalgebra of A. This is shown as follows. Let m_1, m_2 be the indices of N_1 and N_2 respectively; $N_3 = N_1 \frown N_2$ is nilpotent and, since $N_3^{m_1} \le N_1^{m_1} = 0$, its index m_3 is not greater than m_1. Now

$$\begin{aligned}(N_1 + N_2)^2 &= N_1^2 + N_2^2 + N_1N_2 + N_2N_1 \\ &\le N_1^2 + N_2^2 + N_3 \le N_1 + N_2\end{aligned}$$

since it follows from the invariance of N_1 and N_2 that N_1N_2 and N_2N_1 are contained in both N_1 and N_2 and therefore in N_3. Similarly

$$(N_1 + N_2)^r \le N_1^r + N_2^r + N_3$$

so that, if m is the greater of m_1 and m_2,

$$(N_1 + N_2)^m \leq N_1^m + N_2^m + N_3 = N_3$$

and hence $N_1 + N_2$ is a nilpotent subalgebra. Further

$$A(N_1 + N_2) = AN_1 + AN_2 \leq N_1 + N_2$$
$$(N_1 + N_2)A = N_1A + N_2A \leq N_1 + N_2$$

so that $N_1 + N_2$ is invariant. It follows that the totality of all nilpotent invariant subalgebras is itself a nilpotent invariant subalgebra; this algebra is called the maximal nilpotent invariant subalgebra or *radical* of A.

An algebra A which is not nilpotent and which has no radical is said to be *semi-simple*; if in addition it has no invariant subalgebra, it is said to be *simple*.[6] We have then the following theorem whose proof we leave to the reader.

THEOREM 3. *If N is the radical of a non-nilpotent algebra A, then $(A - N)$ is semi-simple.*

10.07 Idempotent elements. In the preceding section we defined a nilpotent algebra of index m as one for which $A^m = 0$, $A^{m-1} \neq 0$. An immediate consequence of this definition is that every element of a nilpotent algebra is nilpotent; we shall now prove the converse by showing that, if A is not nilpotent, it contains an idempotent element.

THEOREM 4. *Every algebra which is not nilpotent contains an idempotent element.*

Let $A = (a_1, a_2, \cdots, a_\alpha)$ be an algebra of order α. If $aA = A$ for some element a in A, then $ax = 0$ only when $x = 0$; for $aA = A$ implies that $aa_1, aa_2, \cdots, aa_\alpha$ is a basis, which means that there is no relation of the form

$$0 = \Sigma\xi_i aa_i = a\Sigma\xi_i a_i$$

except when every $\xi_i = 0$. Also, if $aA = A$, there must be an element e in A such that $ae = a$; this gives $ae^2 = ae$ or $a(e^2 - e) = 0$ and hence $e^2 = e$.

The theorem is true of algebras of order 1; assume it true for algebras of order less than α. If $a_iA = A$ for some a_i, the theorem has just been shown to hold. If $a_iA < A$ for every a_i in the basis of A, then, since $(a_iA)^2 = a_iAa_iA \leq a_iA$, either a_iA contains an idempotent element or, being of order less than α, it is nilpotent. Now $(Aa_iA)^r \leq A(a_iA)^r$ and therefore Aa_iA is also nilpotent; but

$$A \cdot Aa_iA \leq Aa_iA, \qquad Aa_iA \cdot A \leq Aa_iA$$

so that Aa_iA is invariant and being nilpotent is contained in the radical N of A. Hence

$$A^3 = \sum_i Aa_iA \leq N$$

[6] Simple algebras are usually excluded from the class of semi-simple algebras; it seems more convenient however to include them.

The statement that A is not nilpotent is made in order to exclude the algebra of order 1 defined by a single element whose square is 0.

so that A^3, and therefore also A, is nilpotent, contrary to the hypothesis of the theorem. It follows that some a_iA is not nilpotent and being of lower order than A contains an idempotent element by assumption. The theorem is therefore proved.

The following lemma is an immediate consequence of Theorem 4.

LEMMA 1. *A non-nilpotent algebra cannot have a basis every element of which is nilpotent, nor a basis for which the trace of every element is* 0.

For, if every element of the basis is nilpotent, the trace of every element of the algebra is 0 whereas the trace of an idempotent element is not 0 since the only roots of its characteristic equation are 0 and 1.

If e is the only idempotent element in eAe, it is said to be *primitive*. An algebra which is not nilpotent contains at least one primitive idempotent element. For, if eAe contains an idempotent element $e_1 \neq e$, then $e_1(e - e_1) = 0$ so that e_1eAee_1 does not contain $e - e_1$ and is therefore of lower order than eAe; since the order of eAe is finite, a succession of such steps must lead to a primitive idempotent element.

THEOREM 4.5. *A simple algebra has a principal unit.*

If A is not nilpotent, it contains an idempotent element e. If a is any element of A, we may set $a = a_1 + a_2$ where

$$a_1 = ea + ae - eae < eA + Ae, \qquad a_2 = a - a_1, \qquad ea_2 = 0 = a_2e.$$

We can therefore find a complex A_1 such that

$$A = eA + Ae + A_1, \qquad eA + Ae \frown A_1 = 0, \qquad eA_1 = 0 = A_1e.$$

If A_1 is not nilpotent, it contains an idempotent element e' and $e + e'$ is also idempotent since $ee' = 0 = e'e$. We can therefore take $e + e'$ in place of e so reducing the order of A_1, and after a finite number of such steps we arrive at a stage at which A_1 contains no idempotent element and is therefore nilpotent; we shall now assume that e was chosen at the start so that A_1 is nilpotent; we shall also assume that e is not an identity for A and there is no real loss of generality in assuming in addition that it is not a left-hand identity.

Let r be the index of A_1. If $r > 1$ and $x \neq 0$ is any element of A_1^{r-1} then $xA_1 = 0 = A_1x$, $ex = 0$; if $r = 1$, then $A_1 = 0$ and since e is not a left-hand identity, $e \leq eA < A$ so that there is an $x \neq 0$ such that $ex = 0$; we have therefore in both cases

$$xA_1 = 0 = A_1x, \qquad ex = 0.$$

We now have $Ax = eAx$, $AxA = eAxA$; hence $Ax < A$, $AxA < A$ and AxA is therefore an invariant subalgebra of A; if $AxA = 0$, then Ax is invariant and not equal to A; if $Ax = 0$, then xA is a proper invariant subalgebra unless it is 0 in which case $X = \{x\}$ is a non-zero invariant subalgebra of A. In the case of a simple algebra it follows that e is an identity.

Corollary. An algebra without a principal unit is not semi-simple. For $(Ax)^2 = AxAx = AxeAx = 0$ if $A_1 \neq 0$.

10.08 Matric subalgebras. Let A be an algebra which contains the identity and let e_1 be a primitive idempotent element; then $e_\alpha = 1 - e_1$ is also idempotent and, if e_iAe_j is denoted by A_{ij}, then

$$A = (e_1 + e_\alpha)A(e_1 + e_\alpha) = A_{11} + A_{1\alpha} + A_{\alpha 1} + A_{\alpha\alpha}.$$

Suppose in the first place that $A_{\alpha 1}A_{1\alpha}$ is not nilpotent; there is then some $a_{12} < A_{1\alpha}$ such that $A_{\alpha 1}a_{12}$, which is an algebra, is not nilpotent since otherwise $A_{\alpha 1}A_{1\alpha}$ would have a basis of nilpotent elements, which is impossible by Lemma 1; hence some such $A_{\alpha 1}a_{12}$ contains an idempotent element, say $e_2 = a_{21}a_{12}$. If e_2 is not primitive in A, say $e_2 = e' + e''$, $e'e'' = 0 = e''e'$, where e' is primitive in A, then $a_{12}e' \neq 0$ since otherwise

$$e' = e_2e' = a_{21}a_{12}e' = 0;$$

also $e' < A_{\alpha\alpha}$ since $0 = e_1e_2 = e_1e' + e_1e''$ so that $e_1e'' = -e_1e'$ and therefore

$$e_1e' = -e_1e''e' = 0$$

and similarly $e'e_1 = 0$; we may therefore take $a'_{12} = a_{12}e'$ and $a'_{21} = e'a_{21}$ in place of a_{12} and a_{21}, which gives e' in place of e_2. We can therefore assume a_{12} so chosen that e_2 is primitive in A; also, since $e_2a_{21}a_{12}e_2 = e_2^3 = e_2$, then, replacing a_{21} by e_2a_{21}, if necessary, we may assume $e_2a_{21} = a_{21}$ and similarly $a_{12}e_2 = a_{12}$.

The element $a_{12}a_{21}$ is not 0 since

$$a_{21}a_{12}a_{21}a_{12} = a_{21}a_{12}\cdot a_{21}a_{12} = e_2^2 = e_2,$$

and it is idempotent since

$$(a_{12}a_{21})^2 = a_{12}\cdot a_{21}a_{12}\cdot a_{21} = a_{12}e_2a_{21} = a_{12}a_{21}.$$

But $a_{12}a_{21} \leq A_{1\alpha}A_{\alpha 1} \leq A_{11}$ and, since e_1 is primitive, it follows that $a_{12}a_{21} = e_1$. For the sake of symmetry we now put $a_{11} = e_1$, $a_{22} = e_2$, and we then have a matric subalgebra of A, namely $a_{11}, a_{12}, a_{21}, a_{22}$.

Since $A_{\alpha 1}(A_{1\alpha}A_{\alpha 1})^rA_{1\alpha} = (A_{\alpha 1}A_{1\alpha})^{r+1}$, it follows that $A_{1\alpha}A_{\alpha 1}$ and $A_{\alpha 1}A_{1\alpha}$ are either both nilpotent or both not nilpotent. Suppose that both are nilpotent; then, since their product in either order is 0, their sum is nilpotent and, because $(A_{1\alpha} + A_{\alpha 1})^2 = A_{1\alpha}A_{\alpha 1} + A_{\alpha 1}A_{1\alpha}$, it follows that

$$N_1 = A_{1\alpha} + A_{\alpha 1} + A_{1\alpha}A_{\alpha 1} + A_{\alpha 1}A_{1\alpha}$$

is nilpotent. Now

$$\begin{aligned} AN_1 &= (A_{11} + A_{1\alpha} + A_{\alpha 1} + A_{\alpha\alpha})\,(A_{1\alpha} + A_{\alpha 1} + A_{1\alpha}A_{\alpha 1} + A_{\alpha 1}A_{1\alpha}) \\ &= A_{11}A_{1\alpha} + A_{11}A_{1\alpha}A_{\alpha 1} + A_{1\alpha}A_{\alpha 1} + A_{1\alpha}A_{\alpha 1}A_{1\alpha} + A_{\alpha 1}A_{1\alpha} \\ &\qquad + A_{\alpha 1}A_{1\alpha}A_{\alpha 1} + A_{\alpha\alpha}A_{\alpha 1} + A_{\alpha\alpha}A_{\alpha 1}A_{1\alpha} \\ &\leq N_1 \end{aligned}$$

since $A_{ij}A_{pq} = 0$ $(p \neq j)$, $A_{ij}A_{jq} \leq A_{iq}$. Similarly $N_1A \leq N_1$. Hence N_1 lies in the radical of A.

Suppose that we have found a matric subalgebra a_{ij} $(i, j = 1, 2, \cdots, r-1)$ such that $e_i = a_{ii}$ $(i = 1, 2, \cdots, r-1)$ are primitive idempotent elements of A; let $e_\alpha = 1 - \sum_1^{r-1} e_i$ and set $A_{ij} = e_iAe_j$ as before. Suppose further that $A_{\alpha i}A_{i\alpha}$ is not nilpotent for some i; we may then take $i = 1$ without loss of generality. By the argument used above there then exists a primitive idempotent element $e_r = a_{rr} < A_{\alpha 1}A_{1\alpha}$ and elements $a_{r1} < A_{\alpha 1}$, $a_{1r} < A_{1\alpha}$ such that

$$a_{r1}a_{1r} = a_{rr}, \qquad a_{1r}a_{r1} = a_{11},$$
$$a_{rr}a_{r1} = a_{r1}, \qquad a_{1r}a_{rr} = a_{1r}.$$

If we set

$$a_{ir} = a_{i1}a_{1r}, \qquad a_{ri} = a_{r1}a_{1i} \qquad (i = 1, 2, \cdots, r-1),$$

then $a_{ir} \neq 0$ since $a_{1i}a_{ir} = a_{1r}$, and a_{ij} $(i, j = 1, 2, \cdots, r)$ form a matric algebra of higher order than before.

Again, if every $A_{\alpha i}A_{i\alpha}$ is nilpotent, it follows as above that each $A_{i\alpha}A_{\alpha i}$ is also nilpotent and hence

$$N_{r-1} = \sum_{i,j=1}^{r-1} (A_{i\alpha} + A_{\alpha i} + A_{i\alpha}A_{\alpha j} + A_{\alpha i}A_{i\alpha}),$$

having a nilpotent basis, is itself nilpotent; and it is readily seen as before that it is invariant and therefore belongs to the radical of A.

We can now treat $A_{\alpha\alpha}$ in the same way as A, and by doing so we derive a set of matric algebras $M_p(a^p_{ij}; i, j = 1, 2, \cdots, r_p)$ with the identity elements

$$a_p = \sum_{i=1}^{r_p} a^p_{ii}$$

such that $\Sigma a_p = 1$; also

$$N' = \sum_{p \neq q} (a_pAa_q + a_pAa_qAa_p)$$

is contained in the radical N of A. We have therefore the following Lemma.

Lemma 2. *If A is an algebra with an identity, there exists a set of matric subalgebras $M_p = (a^p_{ij}; i, j = 1, 2, \cdots, r_p)$ with the principal units*

$$a_p = \sum_{i=1}^{r_p} a^p_{ii} \qquad (p = 1, 2, \cdots, k)$$

such that $a_pa_q = 0$ $(p \neq q)$ and $\Sigma a_p = 1$, and such that

$$N' = \sum_{p \neq q} (a_pAa_q + a_pAa_qAa_p)$$

lies in the radical N of A. Further each a^p_{ii} is a primitive idempotent element of A.

Corollary. $B_k = a_k A a_k + N'$ *is an invariant subalgebra of* A. For

$$AB_k = \Sigma a_i A(a_k A a_k + \sum_{p \neq q} a_p A a_q + a_p A a_q A a_p)$$
$$= a_k A a_k + N' = B_k.$$

10.09 We shall now consider the properties of the algebras $a_p A a_p$ where a_p $(p = 1, 2, \cdots, k)$ are the idempotent elements defined in Lemma 2.

Lemma 3. $a_p A a_p$ *is the direct product of* M_p *and an algebra* B_p *in which the principal unit is the only idempotent element.*

The first part of this lemma is merely a particular case of Theorem 2. That B_p contains only one idempotent element is seen as follows. If e is a primitive idempotent element of B_p, then $a_{11}^p e$ and $a_{11}^p(a_p - e)$ are distinct and, if not zero, are idempotent and lie in $a_{11}^p A a_{11}^p$; but this algebra contains only one idempotent element since a_{11}^p is primitive; hence $a_{11}^p(a_p - e) = 0$, and therefore $e = a_p$ is the only idempotent element in B_p.

Lemma 4. *If* B *is an algebra whose principal unit* 1 *is its only idempotent element, any element of* B *which is singular*[7] *is nilpotent; and the totality of such elements forms the radical of* B.

The proof of the first statement is immediate; for, if a is singular, the algebra $\{a\}$ generated by a does not contain the principal unit and, since B contains no other idempotent element, a is nilpotent by Theorem 4. To prove the second part, let x and y be nilpotent but $z = x + y$ non-singular; then $1 = z^{-1}x + z^{-1}y = x_1 + y_1$. Here x_1 and y_1 are singular and therefore nilpotent. If m is the index of x_1, then

$$(1 - x_1)(1 + x_1 + x_1^2 + \cdots + x_1^{m-1}) = 1$$

and this is impossible since $y_1 = 1 - x_1$ is nilpotent. Hence z is also nilpotent and the totality of nilpotent elements forms an algebra; and this algebra is invariant since the product of any element of B into a nilpotent element is singular and therefore nilpotent. It follows that B is a division algebra whenever it has no radical, that is, when it is semi-simple.

10.10 **The classification of algebras.** We shall now prove the main theorem regarding the classification of algebras in a given field F.

Theorem 5. (i) *Any algebra which contains an identity can be expressed in the form*

(14) $$A = S + N$$

[7] An element of B is *singular in* B if it does not have an inverse relatively to the principal idempotent element of B.

where N is the radical of A and S is a semi-simple subalgebra; S is not necessarily unique but any two determinations of it are simply isomorphic.

(ii) *A semi-simple algebra can be expressed uniquely as the direct sum of simple algebras.*

(iii) *A simple algebra can be expressed as the direct product of a division algebra D and a simple matric algebra M; these are not necessarily unique but, if D_1, M_1, D_2, M_2 are any two determinations of D and M, then $D_1 \simeq D_2$, $M_1 \simeq M_2$.*

We have seen in Lemma 2 that $A = \Sigma a_p A a_p + N'$, where $N' \leq N$, and also in Lemmas 3, 4 that $a_p A a_p = M_p \times B_p$, where M_p is a simple matric algebra. The first part of the theorem therefore follows for A when it is proved for any algebra like B_p and when it is shown that the direct product of M_p by a division algebra is simple; for, if $B_p = D_p + N_p$, then D_p is a division algebra and

$$a_p A a_p = M_p \times D_p + M_p \times N_p, \qquad M_p N_p \leq N.$$

If the field F is one in which every equation has a root, the field itself is clearly the only division algebra and hence $M_p D_p = M_p$; in this case part (i) is already proved. Further, the theorem is trivial for algebras of order 1; we may, therefore, as a basis for a proof by induction assume it is true for algebras of order less than the order α of A.

If the field F is extended to $F(\xi)$ by the adjunction of an algebraic irrationality ξ of degree $\rho + 1$, we get in place of A an algebra $A' = A(\xi)$ which has the same basis as A but which contains elements whose coordinates lie in $F(\xi)$ but not necessarily in F; all elements of A are also elements of A'. Regarding A' we have the following important lemma.

LEMMA 5. *If N is the radical of A, the radical of $A' = A(\xi)$ is $N' = N(\xi)$.*

Let $A = C + N$, $C \frown N = 0$, and let the radical of A' be N''; then clearly $N'' \geq N'$. If $N'' > N'$, there is an element of N'' of the form

$$c'' = c_0 + c_1\xi + \cdots + c_\rho \xi^\rho, \qquad (c_i < C,\ c_0 \neq 0).$$

Since c'' is nilpotent,

$$0 = \mathrm{tr}(c'') = \mathrm{tr}(c_0) + \xi \mathrm{tr}(c_1) + \cdots$$

and since $\mathrm{tr}(c_0)$, $\mathrm{tr}(c_1)$, $\cdots$ are rational in F, each is separately 0. But, if a_1, a_2 are arbitrary elements in A,

$$a_1 c'' a_2 = a_1 c_0 a_2 + a_1 c_1 a_2 \xi + \cdots$$

lies in N'' and, since each $a_1 c_i a_2$ is rational in F, the trace of each is 0 as above. Hence the trace of every element in Ac_0A is 0 from which it follows by Lemma 1 that Ac_0A is nilpotent and being invariant and also rational it must lie in N (cf. §10.06). But Ac_0A contains c_0 since A contains 1 whereas $C \frown N = 0$; hence no elements of N'' such as c'' exist and the lemma is therefore true.

We may also note that, if B, C are complexes for which $B \frown C = 0$, and B', C' the corresponding complexes in A', then also $B' \frown C' = 0$.

Suppose now that the identity is the only idempotent element of A and that the first part of the theorem is true for algebras of order less than α. Let $a \neq 1$ be an element of A corresponding to an element $\bar{a}$ of $(A - N)$ and let $f(\lambda)$ be the reduced characteristic function of $\bar{a}$; $f(\lambda)$ is irreducible in F since $(A - N)$ is a division algebra. Since $f(\bar{a}) = 0$, it follows that $f(a) < N$ and hence, if r is the index of $f(a)$, the reduced characteristic function of a is $[f(\lambda)]^r$. If we adjoin to F a root ξ of $f(\lambda)$, this polynomial becomes reducible so that in $A' = A(\xi)$ the difference algebra $(A' - N')$ is no longer a division algebra though by Lemma 5 it is still semi-simple. If we now carry out in $F(\xi)$ the reduction given in Lemma 2, say

$$A' = \Sigma e_p A' e_p + N^*,$$

either the algebras $e_p A' e_p$ are all of lower order than α, or, if $A' = e_1 A' e_1$, then it contains a matric algebra M' of order n^2 $(n > 1)$ and, if we set $A' = M'B'$, as previously, B' is of lower order than α. In all cases, therefore, part (i) of the theorem follows for algebras in $F(\xi)$ of order α when it is true for algebras of order less than α, and its truth in that case is assumed under the hypothesis of the induction.

We may now assume

$$A = C + N, \qquad C \frown N = 0,$$
$$A' = S' + N', \qquad S' \frown N' = 0,$$

where S' is an algebra simply isomorphic with $(A' - N')$; N' has a rational basis, namely that of N (cf. Lemma 5).

If $c_1, c_2, \cdots$ is a basis of C then, since A is contained in A' we have

$$c_i = s_i' + m_i', \qquad s_i' < S', \qquad m_i' < N', \qquad (i = 1, 2, \cdots)$$

and, since $C \frown N = 0$ implies $C' \frown N' = 0$, it follows that $s_1', s_2', \cdots$ form a basis of S', that is, we may choose a basis for S' in which the elements have the form

$$c_i + n_{i0} + n_{i1}\xi + \cdots \qquad (c_i < C,\ n_{ij} < N)$$

where c_i, n_{i0}, $\cdots$ are rational in F. Moreover, since C is only determined modulo N, we may suppose it modified so that n_{i0} is absorbed in c_i; we then have a basis for S'

(15) $$s_i' = c_i + n_{i1}\xi + \cdots + n_{ip}\xi^p = c_i + n_i'.$$

When the basis is so chosen, the law of multiplication in S', say

(16) $$s_i' s_j' = \Sigma \sigma_{ijk} s_k'$$

has constants σ_{ijk} which are rational in F; for $s_i' = c_i \bmod N'$ and c_i is rational. If we now replace s_k' in (16) by its value from (15) and expand, we have

$$c_i c_j + c_i n_j' + n_i' c_j + n_i' n_j' = \Sigma \sigma_{ijk} c_k + \Sigma \sigma_{ijk} n_k',$$

but $n'_i n'_j < (N')^2$ and therefore

$$c_i c_j + c_i n'_j + n'_i c_j = \Sigma \sigma_{ijk} c_k + \Sigma \sigma_{ijk} n'_k \qquad \text{mod } (N')^2,$$

a relation which is only possible if the coefficients of corresponding powers of ξ are also equivalent modulo $(N')^2$ and in particular

$$c_i c_j = \Sigma \sigma_{ijk} c_k \qquad \text{mod } (N')^2.$$

Consequently the algebra A_1 generated by c_i $(i = 1, 2, \cdots, \sigma)$ contains no element of N which is not also in N^2 and hence, except in the trivial case in which $N = 0$, the order of A_1 is less than α. By hypothesis we can therefore choose C rationally in such a way that $c_i c_j = \Sigma \sigma_{ijk} c_k$, that is, such that C is an algebra; part (i) of the theorem therefore follows by induction.

10.11 For the proof of part (ii) we require the following lemmas.

LEMMA 6. *If A contains the identity* 1 *and if B is an invariant subalgebra which has a principal unit e, then*

$$A = B \oplus (1 - e)A(1 - e).$$

Since e is the principal unit of B, which is invariant, $eAe = B$; also $eA(1 - e)$ and $(1 - e)Ae$ are both 0 since Ae and eA lie in B and, if b is any element of B, then $(1 - e)b = b - b = 0$, $b(1 - e) = b - b = 0$; hence

$$(17) \quad A = eAe + (1 - e)\,A(1 - e), \qquad eAe \frown (1 - e)\,A(1 - e) = 0.$$

Further $eAe \cdot (1 - e)\,A(1 - e) = 0 = (1 - e)\,A(1 - e) \cdot eAe$, so that the sum in (17) is a direct sum.

LEMMA 7. *Every invariant subalgebra B of a semi-simple algebra A is semi-simple and therefore contains a principal unit.*

Suppose that B has a radical N; then

$$AN \leq B, \qquad (AN)^2 = ANA \cdot N \leq BN \leq N$$

so that AN is nilpotent. But, since $A^2 = A$, we have $(ANA)^r = (AN)^r A$; hence ANA is a nilpotent invariant subalgebra of A which, since A contains an identity, is not 0 unless N is 0. But A has no radical; hence $N = 0$ and B also has no radical.

In consequence of these lemmas a simple algebra is irreducible and a semi-simple algebra which is not also simple can be expressed as the direct sum of simple algebras. Let

$$A = B_1 \oplus B_2 \oplus \cdots \oplus B_p = C_1 \oplus C_2 \oplus \cdots \oplus C_q$$

be two expressions of A as the direct sum of simple algebras and let the principal units of B_i and C_i be b_i and c_i respectively; then $1 = \Sigma b_i = \Sigma c_i$. We then have $C_k \leq \Sigma b_i C_k b_j \leq C_k$ and therefore

$$C_k = \Sigma b_i C_k b_j = \Sigma b_i C_k b_i$$

since when $i \neq j$ then $b_iC_kb_j \leq B_i \frown B_j = 0$ and $b_iC_kb_i \cdot b_jC_kb_j = 0$. If $b_iC_kb_i \neq 0$, it is an invariant subalgebra of C_k and, since the latter is simple, we have $b_iC_kb_i = C_k$ for this value of i and all other $b_jC_kb_j$ equal 0, and therefore $C_k = b_iAb_i = B_i$. The second part of the theorem is therefore proved.

10.12 We shall now prove part (iii) of Theorem 5 in two stages.

LEMMA 8. *If D is a division algebra and M the matric algebra (e_{ij}; $i, j = 1, 2, \cdots, m$), and if $D \times M = DM$, then DM is simple.*

Let B be a proper invariant subalgebra of $A = DM$. If x is an element of B, then there exists an element y of A such that $xy = 0$, since otherwise we should have $B \geq xA = A$, in other words, every element of B is singular in A and hence $B \frown D = 0$. But

$$x = \Sigma d_{ij}e_{ij}, \qquad d_{ij} < D$$

and $d_{ij} = \sum_p e_{pi}xe_{jp}$ and is therefore contained in B as well as in D. Since $B \frown D = 0$, every $d_{ij} = 0$, that is, $x = 0$ so that $B = 0$. It follows immediately from Lemma 2 that a simple algebra always has the form $D \times M$, and also that $D \simeq e_{ii}Ae_{ii}$.

LEMMA 9. *In a simple algebra all primitive idempotent elements are similar.*

Let e and a be primitive idempotent elements of a simple algebra A. We can then find a matric algebra $M = (e_{ij})$ for which $e_{11} = e$ and such that $A = D \times M$, where D is a division algebra. If $ea = 0 = ae$, we can at the same time choose $e_{22} = a$; and $e_{22} = ue_{11}u^{-1}$ where

$$u = 1 - e_{11} - e_{22} + e_{12} + e_{21} = u^{-1},$$

so that the lemma is true in this case, and we may therefore assume that, say, $ea \neq 0$.

Suppose now that $eae \neq 0$. Since $A = D \times M$, we can express a in the form $\Sigma\alpha_{ij}e_{ij}$ ($\alpha_{ij} < D$), where $\alpha_{11} \neq 0$ since $eae = \alpha_{11}e_{11}$. We have then

$$(ea)^2 = (e_{11}a)^2 = (\alpha_{11}e_{11} + \alpha_{12}e_{12} + \cdots)^2 = \alpha_{11}e_{11}a$$

and hence $b = \alpha_{11}^{-1}ea$ is idempotent. We then have

$$eb = b, \qquad be = e, \qquad ba = b;$$

also $ab = aba = abab$ and, since a is primitive, either $aba = a$ or $aba = 0$; but

$$eabe = eae \neq 0,$$

hence $ab = a$. We then have

$$\begin{aligned} b &= ueu^{-1}, & u &= 1 - b + e, & u^{-1} &= 1 + b - e, \\ b &= v^{-1}av, & v &= 1 - b + a, & v^{-1} &= 1 + b - a, \end{aligned}$$

and hence a and e are similar in this case also.

If $eae = 0$ but $aea \neq 0$, interchanging the rôles of e and a leads to results similar to those just obtained; we can therefore assume $eae = 0 = aea$. If

$$u = 1 + e - ea + ae = 2(2 - e + ea - ae)^{-1},$$

then $uau^{-1} = a - ae$; we can therefore assume $ae = 0$. If

$$v = (1 + e - 2ea) = 2(2 - e + 2ea)^{-1},$$

then $vav^{-1} = a - ea$; we can therefore also assume $ea = 0$, which brings us back to the first case which we considered. The lemma is therefore proved.

Part (iii) of the theorem follows immediately. For, if e and a are primitive idempotent elements of M_1 and M_2 respectively, we can now find w such that $a = wew^{-1}$; but $D_1 \simeq eAe$ and $D_2 \simeq aAa = weAew^{-1}$, which is similar to eAe and therefore to D_1.

10.13 Semi-invariant subalgebras. If B is a subalgebra of A which is such that $AB \leq B$ ($BA \leq B$), it is called a right (left) semi-invariant subalgebra. We shall treat only the case in which A is semi-simple; it has then an identity and if we restrict ourselves, as we shall, to the case of right semi-invariant subalgebras, we may assume $AB = B$.

It is clear that, if $A = A_1 \oplus A_2$, then also $B = B_1 \oplus B_2$, where $A_iB_i = B_i, A_iB_j = 0\ (i \neq j)$. It is sufficient then at first to consider only simple algebras, and in this case we have the added condition that $ABA = A$; that is, we have simultaneously

$$AB = B, \qquad ABA = A. \tag{18}$$

If we call B minimal when it contains no other semi-invariant subalgebra, we have

LEMMA 10. *A minimal right semi-invariant subalgebra of a simple algebra A has the form Au, where u is a primitive idempotent element of A. Conversely, if $u^2 = u$ is primitive, Au is a minimal right semi-invariant subalgebra.*

Let $AC = C$; if $c_1 \neq 0$ is any element of C, and $C_1 = Ac_1 \leq C$, then $AC_1 = C_1$. Suppose $C_1 < C$; then in the same way if c_2 is any element of C_1, we have $C_2 = Ac_2 \leq C_1$. If $C_2 < C_1$, we may continue this process and after a finite number of steps we shall arrive at an algebra $B \neq 0$ such that $Ab = B$ for every element b of B which is not 0. Since A is simple, $AbA = A$ and $B^2 = B$, so that B contains a primitive idempotent element u and $Au = B$. If u is not also primitive in A, let $u = u_1 + u_2, u_iu_j = 0\ (i \neq j), u_i^2 = u_i \neq 0$. Then $u_1u = u_1$ so that u_1 is in B; hence u must be primitive in A, if it is so in B.

Since $B = Au$, every x in B has the form au and hence $xu = x$. But also $B = Ax$ and, from the manner in which B was chosen, either $Bx = 0$ or $Bx = B$. If $Bx = 0$, then $ux = 0$ and therefore

$$x^2 = xu \cdot xu = 0.$$

Also, if x is nilpotent, then $x^2 = 0 = ux$; for $uAu = uBu$ is simple since, by the proof of Lemma 4, it is a division algebra, and $ux = uxu < uBu$. If $Bx = B$, then there is a unique b such that $bx = x$ and, since b is then idempotent, we have $ux = x$, that is, x lies in uAu. If $B = Au$, then $AB = A^2u \leq B$ so that B is a right semi-invariant subalgebra of A. If C is minimal, then $B = C$ as desired.

Conversely, let $B = Au$, u primitive; then the only idempotent quantity of B has been shown above to be u and, if B were not primitive, we should have $B > C = Av$, v primitive, which is impossible.

Suppose now that B is not minimal and let $e_1, e_2, \cdots, e_r$ be a complete set of primitive supplementary idempotent elements in B. Then $B_r = Ae_1 + Ae_2 + \cdots + Ae_r$ is semi-invariant in A. Let b be an element of B which is not in B_r; since $b \neq \Sigma be_i$, we may replace b by $b - \Sigma be_i$ and so assume every $be_i = 0$ in which case clearly $Ab \frown B_r = 0$. But, if $b \neq 0$, then Ab contains an idempotent element e such that $e_ie = 0$ $(i = 1, 2, \cdots, r)$ and $e_{r+1} = e - \Sigma e_ie$ is an idempotent element supplementary to the given complete set, which is impossible. We therefore have the following theorem.

Theorem 6. *If A is simple and $AB = B$ is a semi-invariant subalgebra, then*

$$B = Ae_1 + Ae_2 + \cdots + Ae_r$$

where $e_1, e_2, \cdots, e_r$ is a complete supplementary set of primitive idempotent elements of B; and these idempotent elements are also primitive in A.

We shall assume that A is semi-simple, say

$$A = S_1 \oplus \cdots \oplus S_t, \tag{19}$$

when each S_i is simple and

$$S_i = D_i \times M_i. \tag{20}$$

As previously (cf. Lemma 2) we may set $M_i = (e^i_{pq})$, $p, q = 1, 2, \cdots, n_i$, where e^i_{pp} form a set of supplementary primitive idempotent elements and $\sum_{i,p} e^i_{pp} = 1$. If B is any invariant subalgebra, then $B = \sum_{i,p} Be^i_{pp}$ and Be^i_{pp} is a right semi-invariant subalgebra; if B is minimal, we have already seen that it has the form Bu where u is a primitive idempotent element, and therefore we have $B = Be^i_{pp} = S_ie^i_{pp}$ for some i and p. If set $B_{ip} = S_ie^i_{pp}$, then

$$B_{ip}e^i_{pq} = S_ie^i_{pp}e^i_{pq} = M_iD_ie^i_{pq} = S_ie^i_{qq} = B_{iq}.$$

We have therefore the following theorem.

Theorem 7. *If A is semi-simple and is given by* (19), *and if e^i_{pp} form a complete* *set of supplementary primitive idempotent elements such that* $\sum_{p=1}^{n_i} e^i_{pp} = u_i$ *is the identity of* S_i, *then every minimal right semi-invariant subalgebra has the form*

$$B_{ip} = S_i e^i_{pp}. \tag{21}$$

Moreover, there is a number e^i_{pq} *in* S_i *such that*

$$B_{ip} e^i_{pq} = B_{iq}. \tag{22}$$

10.14 **The representation of a semi-simple algebra.** Let A be a linear associative algebra over F with the identity 1, and designate elements of A by a. A *representation* of A is a set, $U(a)$, of matrices of order n such that $a \to U(a)$ is a correspondence between the elements of A and the matrices of the set in which the following conditions are satisfied

$$U(1) = 1_n, \qquad U(a + b) = U(a) + U(b), \qquad U(ab) = U(a)\,U(b), \qquad U(\alpha a) = \alpha U(a) \tag{23}$$

for every a and b of A and every scalar α in F.

We can now, as in chapter I, associate with the matrices $U(a)$ a vector space R with a given fundamental basis, and a change of basis corresponds to replacing $U(a)$ by $PU(a)P^{-1}$, an equivalent representation (cf. 1.08). A subspace R_1 of R is invariant under A (cf. 5.16) if every matrix $U(a)$ carries each vector of R_1 into a vector of R_1. If $R_1 \neq 0$, we may set $R = R_1 + R_2$ $(R_1 \frown R_2 = 0)$; and since we are only interested in the equivalence of representations, we may suppose the basis R so chosen that

$$U(a) = \left\| \begin{matrix} U_1(a) & U_3(a) \\ 0 & U_2(a) \end{matrix} \right\|. \tag{24}$$

The representation is said to be *reducible* in this case, and it is evident that both $U_1(a)$ and $U_2(a)$ give representations of A.

If R has no proper invariant subspace, then $U(A)$ and R are said to be *irreducible*. It is now clear that we may write

$$R = R_1 + R_2 + \cdots + R_s$$

where $R_t = R_1 + \cdots + R_t$ is the invariant subspace of least order which contains R_{t-1}, $(R_0 = 0)$, and in this case

$$U(a) = \left\| \begin{matrix} U_{11}(a) & U_{12}(a) & \cdots & U_{1s}(a) \\ 0 & U_{22}(a) & \cdots & U_{2s}(a) \\ \cdot & \cdot & \cdots & \cdot \\ \cdot & \cdot & \cdots & \cdot \\ 0 & 0 & \cdots & U_{ss}(a) \end{matrix} \right\| \tag{25}$$

and the representations $U_1(a), \cdots, U_s(a)$ are irreducible. If in addition $R_2, \cdots, R_t$ are themselves invariant for some t, then $U_{ij}(a) = 0$ $(i \neq j;\ i, j = 1, 2, \cdots, t)$, and we say that $U(a)$ is *decomposable*.

A particular case of fundamental importance arises when we take R to be A itself, that is, if x is a variable element of A, then $x' = ax$ corresponds to a linear transformation in the basis of R (or A), say

$$x' = ax = U(a)(x),$$

and $U(a)$ has the property given in (23) and so is a representation of A. It is obviously the representation of (6) and is one-to-one; it is called the *regular* representation.

The invariant subspaces of A are evidently its right semi-invariant subalgebras B. If $e_1, e_2, \cdots, e_t$ is a basis of B and

$$ae_i = \Sigma\alpha_{ji}e_j, \tag{26}$$

then the matrices $U(a) = \| \alpha_{ij} \|$ give a representation of A on the subspace B. Suppose now that $V(a)$ is a given representation, R the corresponding subspace, and B a right semi-invariant subalgebra of A. If y is any vector of R, then the set of vectors of the form $V(b)(y)$ is an invariant subspace of R, since

$$V(a)V(b) = V(ab) = V(b_a), \qquad b_a < B. \tag{27}$$

From (27) it is seen immediately that the set B' of elements b' in B for which $V(b')y = 0$ forms a right semi-invariant subalgebra of B and hence, if B is minimal, either $B' = 0$ or $B' = B$. If $B' = 0$, then $V(e_1)y, \cdots, V(e_t)y$ is a basis of the set $(V(b)y)$ and

$$V(a)V(e_i)y = V(ae_i)y = \Sigma\alpha_{ji}V(e_j)y.$$

But then the vectors of the form $V(b)y$ give a representation of A equivalent to that determined by B in (26).

We shall now prove the following theorem.

THEOREM 8. *If the regular representation of an algebra is decomposable, then every representation is decomposable and its irreducible components are contained in the regular representation.*

Suppose that the regular representation of A is decomposable; then $A = B_1 + B_2 + \cdots + B_s$, where the B_j are irreducible equivalent subspaces of A, that is, minimal semi-invariant subalgebras such that $B_j \frown B_k = 0$ for $j \neq k$. Let $y_1, y_2, \cdots, y_n$ be a basis of the space R of a representation of A. Since A has an identity, we have

$$\begin{aligned} R = AR &= B_1R + B_2R + \cdots + B_sR \\ &= B_1y_1 + B_2y_2 + \cdots + B_2y_1 + \cdots + B_sy_n. \end{aligned} \tag{28}$$

As we have seen above, if $B_ky_j \neq 0$, it is a subspace of R which gives a representation equivalent to that given by B_k; it follows that either $B_ky_j = 0$ or it is an invariant subspace of R.

The intersection of the invariant subspaces is also invariant so that either

$B_k y_j \frown B_p y_q = 0$ or $B_k y_j = B_p y_q$; hence we may select from the spaces $B_k y_q$ in (28) a set of independent irreducible invariant subspaces determining R. This proves Theorem 8.

Consider now a semi-simple algebra

$$A = S_1 \oplus \cdots \oplus S_r$$

where S_i is a simple algebra. We may write

$$1 = \Sigma u_{ij} \qquad (i = 1, \cdots, r;\ j = 1, \cdots, n_i)$$

where the u_{ij} form a complete set of supplementary primitive idempotent elements of A. Then

$$A = \Sigma A u_{ij} = \Sigma B_{ij}$$

where $B_{ij} = Au_{ij}$ is a minimal right invariant subalgebra of A. We have then decomposed A into irreducible invariant subspaces and have proved the first part of the following theorem.

THEOREM 9. *The regular representation of a semi-simple algebra is decomposable, and its reducible components are those obtained by the use of the B_{ij} as representation spaces. The representations given by any pair B_{ij}, B_{ik} are equivalent while B_{ij}, B_{tk} give inequivalent representations for $j \neq k$.*

For by Theorem 7 we have $B_{ij}e^i_{jk} = B_{ik}$ so that the proof of Theorem 7 with $y = e^i_{jk}$ shows that the representation by B_{ij} is equivalent to that by B_{ip}. In the representation by B_{ij} we have

$$e_i = \sum_{j=1}^{n} u_{ij} \rightarrow 1_{n_i},$$

where 1_{n_i} is an identity matrix corresponding to the identity transformation on B_{ij} since e_i is the principal unit of B_{ij}. But in the representation by B_{tk}, we have $e_i \rightarrow 0$. Evidently these representations cannot be similar.

10.15 Group algebras. If $\mathfrak{G} = (g_1 = 1, g_2, \cdots, g_m)$ is a finite group, the group relation $g_i g_j = g_{ij}$ is a particular case of the associative product defined in (2) and, when it is used in conjunction with addition, we get an associative algebra G of which $(g_1, g_2, \cdots, g_m)$ is a basis and g_1 the identity.

The representation of $\mathfrak{G}$ as a regular permutation group

$$h_i = \begin{pmatrix} g_1 & g_2 & \cdots & g_m \\ g_{i1} & g_{i2} & \cdots & g_{im} \end{pmatrix}$$

corresponds to the representation of G as a set of matrices, the matrix h_i being

$$h_i = \sum_{p=1}^{m} e_{ip\,p} \qquad (g_i g_p = g_{ip}).$$

Since $i_p = p$, that is, $g_i g_p = g_p$, only when g_i is the identity, the matrix h_i has no coordinate in the main diagonal except for $i = 1$ in which case h_1 is the identity matrix; hence

$$\text{(29)} \qquad \operatorname{tr}(h_1) = m, \qquad \operatorname{tr}(h_i) = 0 \qquad (i \neq 1).$$

It follows from this that G is semi-simple. For if $u = \Sigma\eta_i h_i$ is the matrix corresponding to some element of the radical N, then $\operatorname{tr}(u) = 0$ since u is nilpotent. If $u \neq 0$, some coordinate, say η_p, is not 0 and in $h_p^{-1}u$, which also corresponds to some element of N, the coefficient of h_1 is not 0; we may therefore assume $\eta_1 \neq 0$ provided $N \neq 0$. But using (18) we get

$$0 = \operatorname{tr}(u) = \Sigma\eta_i \operatorname{tr}(h_i) = m\eta_1;$$

hence the assumption that $u \neq 0$ leads to a contradiction and therefore $N = 0$, that is, G is semi-simple. This gives the following theorem.

THEOREM 10. *A group algebra is semi-simple. It is therefore the direct sum of simple algebras and, if the field of the coefficients is sufficiently extended, it is the direct sum of simple matric algebras.*

The whole of the representation theory developed in the previous section can now be applied to groups.

APPENDIX I

NOTES[1]

Chapter I

The calculus of matrices was first used in 1853 by Hamilton (1, p. 559ff, 480ff) under the name of "Linear and vector functions." Cayley used the term *matrix* in 1854, but merely for a scheme of coefficients, and not in connection with a calculus. In 1858 (2) he developed the basic notions of the algebra of matrices without recognizing the relation of his work to that of Hamilton; in some cases (e.g., the theory of the characteristic equation) Cayley gave merely a verification, whereas Hamilton had already used methods in three and four dimensions which extend immediately to any number of dimensions. The algebra of matrices was rediscovered by Laguerre (9) in 1867, and by Frobenius (18) in 1878.

1.03 Matric units seem to have been first used by B. Peirce (17); see also Grassmann (5, §381).

1.10 For the history of the notion of rank and nullity see Muir, *Theory of Determinants*, London 1906–1930; the most important paper is by Frobenius (290).

Chapter II

2.01–03 The principle of substitution given in §2.01 was understood by most of the early writers, but was first clearly stated by Frobenius, who was also the first to use the division transformation freely (20, p. 203).

2.04 The remainder theorem is implicit in Hamilton's proof of the characteristic equation; see also Frobenius (280).

2.05–12 The characteristic equation was proved by general methods for $n = 3, 4$ by Hamilton (1, p. 567; 8, p. 484ff; cf. also 4, 6). The first general statement was given by Cayley (2); the first general proof by Frobenius (18). See also the work of Frobenius cited below and 9, 10, 39, 41, 56, 59.

Hamilton, Cayley and other writers were aware that a matrix might satisfy an equation of lower degree than n, but the theory of the reduced equation seems to be due entirely to Frobenius (18, 140).

The theory of invariant vectors was foreshadowed by Hamilton, but the general case was first handled by Grassmann (5).

2.10 See Sylvester (42, 44) and Taber (96); see also 252.

2.13 The square root of a matrix was considered by Cayley (3, 12), Frobenius (139) and many others.

Chapter III

3.01 The idea of an elementary transformation seems to be due in the main to Grassmann (5).

---

[1] In these Notes, numbers refer to the Bibliography unless otherwise indicated.

3.02–07 The theory of pairs of bilinear forms, which is equivalent to that of linear polynomials, was first given in satisfactory form by Weierstrass (see Muth, 175) although the importance of some of the invariants had been previously recognized by Sylvester. The theory in its matrix form is principally due to Frobenius (18, 20).

The theory of matrices with integral elements was first investigated by Smith (see Muth, 175) but was first given in satisfactory form by Frobenius (20). The form given in the text is essentially that of Kronecker (92).

3.04 The proof of Theorem 3 is a slight modification of that of Frobenius (20).

3.08 Invariant vectors were discussed by Hamilton (1, 8) and other writers on quaternions and vector analysis. The earliest satisfactory account seems to be that of Grassmann (5).

Chapter IV

The developments of this chapter are, in the main, a translation of Kronecker's work (see Muth, 175, p. 93ff). See also de Séguier (259).

Chapter V

5.03 From the point of view of matrix theory, the principal references are Schur (198), Rados (105, 106), Stephanos (185), and Hurwitz (117). See Loewy (284, p. 138) for additional references; also Muir, *Theory of Determinants*, London 1906–1930.

5.09 Non-commutative determinants were first considered by Cayley (Phil. Mag. 26 (1845), 141–145); see also Joly (195) and Sylvester (43).

5.10–11 See Loewy (284, p. 149); also 176, 178, 185, 198.

5.12 The principal references are Schur (198) and Weyl (440, chap. 5).

Chapter VI

For general references see Loewy (284, pp. 118–137), also Muth (175), Hilton (314, chap. 6, 8) and Muir, *Theory of Determinants*, London 1906–1930.

6.01 The method of proving that the roots are real is essentially that of Tait (10, chap. 5); see also 36, 60, 228, 399.

6.03 See Loewy (284, pp. 130–137), Baker (215) and Frobenius (292). See also 7, 18, 99, 113, 114, 115, 124, 135, 139, 210, 221, 273, 302, 307, 320, 371, 400, 414, 466, 476.

6.04 See Dickson (392).

6.05 See Loewy (284, pp. 128–135).

6.07 For references see Muth (175, p. 125) and Frobenius (139).

Chapter VII

7.10–02 See Cayley (2), Frobenius (18), Bucheim (59), Taber (98, 112), and Hilton (314, chap. 5); also 83, 86, 98, 137, 184, 197, 209, 223, 242, 250, 264, 301, 382.

7.03 See Frobenius (280).

7.05 See Frobenius (140); also 350.

7.06–07 See Sylvester (42, 44) and Taber (96); see also 252.

Chapter VIII

8.01–03 See Sylvester (36), Bucheim (59, 69); also 134, 371.

8.02,07 See Hamilton (1, p. 545ff; 8, §316), Grassmann (5, §454), Laguerre (9). Many writers define the exponential and trigonometric functions and consider the question of convergence, e.g., 79, 80, 103, 389, 449; also in connection with differential equations, 13, 133, 258.

8.04–05 Roots of 0 and 1 have been considered by a large number of writers; see particularly the suite of papers by Sylvester in 1882–84; also 18, 67, 76, 107, 242, 255, 264, 277, 279, 381, 411, 430, 474, 539.

8.08 See 20, 94, 246, 256, 257, 274, 303, 338, 399.

8.09–11 The absolute value of a matrix was first considered by Peano (75) in a somewhat different form from that given here; see also 273, 348, 389, 472, 473, 494. For infinite products see 133, 324, 326, 389, 494.

8.12 In addition to the references already given above, see 10, 16, 18, 187, 418, 419, and also many writers on differential equations.

Chapter IX

The problem of the automorphic transformation in matrices was first considered by Cayley (3, 7) who, following a method used by Hermite, gave the solution for symmetric and skew matrices; his solution was put in simpler form by Frobenius (18). Cayley failed to impose necessary conditions in the general case which was first solved by Voss (85, 108, 162, 163). The properties of the principal elements were given by Taber (125, 134; see also 127, 149, 156, 158, 231). Other references will be found in Loewy (284, pp. 130–137); see also 9, 19, 153, 154, 161, 167, 168, 169, 187, 229, 371.

APPENDIX II

BIBLIOGRAPHY

1853

1. HAMILTON, W. R.: Lectures on quaternions, p. 480ff. Dublin.

1858

2. CAYLEY, A.: A memoir on the theory of matrices, (1857). Lond. Phil. Trans. 148, 17–37; Coll. Works 2, 475–496.
3. CAYLEY, A.: A memoir on the automorphic transformation of a bipartite quadric function, (1857). Lond. Phil. Trans. 148, 39–46; Coll. Works 2, 497–505.

1862

4. HAMILTON, W. R.: On the existence of a symbolic and biquadratic equation which is satisfied by the symbol of linear or distributive operation on a quaternion. Phil. Mag. (4) 24, 127–128.
5. GRASSMANN, H.: Ausdehnungslehre, p. 241ff. Berlin; Werke 1, Teil 2, p. 240.

1864

6. HAMILTON, W. R.: On the existence of a symbolic and biquadratic equation ..., (1862). Proc. Roy. Irish Acad. 8, 190–191.

1866

7. CAYLEY, A.: A supplementary memoir on the theory of matrices, (1865). Lond. Phil. Trans. 156, 25–35; Coll. Works 5, 438–448.
8. HAMILTON, W. R.: Elements of quaternions, §347ff. London. 2nd ed. 1 (1899), 2 (1901); German translation, 2 vols. (Glan), 1881–1901.

1867

9. LAGUERRE, E. N.: Sur le calcul des systèmes linéaires. Journ. Ec. Polyt. 42, 215–264; Oeuvres 1, 221–267.
10. TAIT, P. G.: An elementary treatise on quaternions, chap. V. Oxford. 2nd ed., Cambridge 1873; 3d ed., Cambridge 1890; German translation (Schreff) 1890; French translation (Plarr) 1882–1884.

1869

11. TAIT, P. G.: Note on the reality of the roots of the symbolical cubic ..., (1867). Proc. Roy. Soc. Edinb. 6, 92–93; Sci. Papers 1, 74–75.

1872

12. CAYLEY, A.: On the extraction of the square root of a matrix of the third order. Proc. Roy. Soc. Edinb. 7, 675–682.
13. TAIT, P. G.: Note on linear differential equations in quaternions, (1870). Proc. Roy. Soc. Edinb. 7, 311–318; Sci. Papers 1, 153–158.
14. TAIT, P. G.: Mathematical notes, (1871). Proc. Roy. Soc. Edinb. 7, 498–499.
15. TAIT, P. G.: Note on the strain function. Proc. Roy. Soc. Edinb. 7, 667–668; Sci. Papers 1, 194–198.

1873

16. TAIT, P. G.: Introduction to quaternions, chap. X (with P. Kelland). London. 2nd ed. 1882; 3d ed. 1904.

1875

17. PEIRCE, B.: On the uses and transformations of linear algebra. Proc. Amer. Acad. Boston (2) 10, 395–400; reprinted, Amer. Journ. Math. 4 (1881), 216–221.

1878

18. FROBENIUS, G.: Über lineare Substitutionen und bilineare Formen, (1877). Crelle 84, 1–63.

1879

19. FROBENIUS, G.: Über die schiefe Invariante einer bilinearen oder quadratischen Form, (1878). Crelle 86, 44–71.
20. FROBENIUS, G.: Theorie der linearen Formen mit ganzen Koefficienten, (1878). Crelle 86, 146–208.
21. PLARR, G.: On the solution of the equation $V\rho\varphi\rho = 0 \cdots$, (1876). Trans. Roy. Soc. Edinb. 28, 45–91.

1880

22. CAYLEY, A.: On the transformation of coordinates. Proc. Cambr. Phil. Soc. 3, 178–184; Coll. Works 11, 136–142.
23. CAYLEY, A.: On the matrix $\begin{matrix}(a\,b)\\ |\,c\,d\,|\end{matrix}$ and in connection therewith the function $\dfrac{ax+b}{cx+d}$. Mess. of Math. 9, 104–109; Coll. Works 11, 252–257.
24. FROBENIUS, G.: Theorie der linearen Formen mit ganzen Koefficienten, (1879). Crelle 88, 96–116.
25. FROBENIUS. G.: Zur Theorie der Transformation der Thetafunktionen, (1879). Crelle, 89, 40–46.

1881

26. LAISANT, C.-A.: Introduction à la méthode des quaternions, chap. 10. Paris.

1882

27. CLIFFORD, W. K.: A fragment on matrices, (1875). Math. Papers, 337–341.
28. PEIRCE, C. S.: On the relative form of quaternions. Johns Hopkins Circ. 1, 179.
29. PLARR, G.: On a particular case of the symbolic cubic, (1881). Proc. Roy. Soc. Edinb. 11, 342–354.
30. SYLVESTER, J. J.: Sur les puissances et les racines de substitutions linéaires. Comptes Rendus 94, 55–59; Math. Papers, 3, 562–564.
31. SYLVESTER, J. J.: Sur les racines des matrices unitaires. Comptes Rendus 94, 396–399; Math. Papers 3, 565–567.
32. SYLVESTER, J. J.: On the properties of a split matrix. Johns Hopkins Circ. 1, 210–211; Math. Papers 3, 645–646.
33. SYLVESTER, J. J.: A word on nonions. Johns Hopkins Circ. 1, 241–242; Math. Papers 3, 647–650.

1883

34. PEIRCE, C. S.: On a class of multiple algebras, (1882). Johns Hopkins Univ. Circ. 2, 3–4.
35. PEIRCE, C. S.: (Correspondence). Johns Hopkins Univ. Circ. 2, 86–88.

36. Sylvester, J. J.: On the equation to the secular inequalities in the planetary theory. Phil. Mag. (5) 16, 267–269.

37. Sylvester, J. J.: On the involution of two matrices of the second order. Brit. Assoc. Report 1883, 430–432; Math. Papers 4, 115–117.

38. Sylvester, J. J.: Sur les quantités formant un groupe de nonions analogues aux quaternions de Hamilton. Comptes Rendus 97, 1336–1340; Math. Papers 4, 118–121.

1884

39. Bucheim, A.: Mathematical notes. Mess. of Math. 13, 62–66.

40. Bucheim, A.: Proof of Professor Sylvester's "Third law of motion." Phil. Mag. (5) 18, 459–460.

41. Forsyth, A. R.: Proof of a theorem by Cayley in regard to matrices. Mess. of Math. 13, 139–142.

42. Sylvester, J. J.: On quaternions, nonions, sedenions etc. Johns Hopkins Circ. 3, 7–9; Math. Papers 4, 122–132.

43. Sylvester, J. J.: On involutants and other allied species of invariants to matrix systems. Johns Hopkins Circ. 3, 9–18, 34–35; Math. Papers 4, 133–145.

44. Sylvester, J. J.: On the three laws of motion in the world of universal algebra. Johns Hopkins Circ. 3, 33–34, 57; Math. Papers 4, 146–151.

45. Sylvester, J. J.: Equations in matrices. Johns Hopkins Circ. 3, 122; Math. Papers 4, 152–153.

46. Sylvester, J. J.: Sur les quantités formant un groupe de nonions analogues aux quaternions de Hamilton. Comptes Rendus 98, 273–276, 471–475; Math. Papers 4, 154–159.

47. Sylvester, J. J.: Sur la solution d'une classe très étendue d'équations en quaternions. Comptes Rendus 98, 651–652; Math. Papers 4, 162.

48. Sylvester, J. J.: Sur une extension de la loi de Harriot relative aux équations algébriques. Comptes Rendus 98, 1026–1030; Math. Papers 4, 169–172.

49. Sylvester, J. J.: Sur les équations monothétiques. Comptes Rendus 99, 13–15; Math. Papers 4, 173–175.

50. Sylvester, J. J.: Sur l'équation en matrices $px = xq$. Comptes Rendus 99, 67–71, 115–116; Math. Papers 4, 176–180.

51. Sylvester, J. J.: Sur la solution du cas le plus général des équations linéaires en quantités binaires .... Comptes Rendus 99, 117–118; Math. Papers 4, 181–182.

52. Sylvester, J. J.: Sur les deux méthodes, celle de Hamilton et celle de l'auteur pour résoudre l'équation linéaire en quaternions. Comptes Rendus 99, 473–476, 502–505; Math. Papers 4, 183–187.

53. Sylvester, J. J.: Sur la solution explicite de l'équation quadratique de Hamilton en quaternions ou en matrices du second ordre. Comptes Rendus 99, 555–558, 621–631; Math. Papers 4, 188–198.

54. Sylvester, J. J.: Sur la résolution générale de l'équation linéaire en matrices d'un ordre quelconque. Comptes Rendus 99, 409–412; 432–436; Math. Papers 4, 199–205.

55. Sylvester, J. J.: Sur l'équation linéaire trinôme en matrices d'un ordre quelconque. Comptes Rendus 99, 527–529; Math. Papers 4, 206–207.

56. Sylvester, J. J.: Lectures on the principles of universal algebra. Amer. Journ. Math. 6, 270–286; Math. Papers 4, 208–224.

57. Sylvester, J. J.: On the solution of a class of equations in quaternions. Phil. Mag. (5) 17, 392–397; Math. Papers 4, 225–230.

58. Sylvester, J. J.: On Hamilton's quadratic equation and the general unilateral equation in matrices. Phil. Mag. (5) 18, 454–458; Math. Papers 4, 231–235.

1885

59. Bucheim, A.: On the theory of matrices, (1884). Proc. Lond. Math. Soc. 16, 63-82.
60. Bucheim, A.: On a theorem relating to symmetrical determinants, (1884). Mess. of Math. 14, 143-144.
61. Bucheim, A.: A theorem on matrices. Mess. of Math. 14, 167-168.
62. Cayley, A.: On the quaternion equation $qQ - Qq' = 0$. Mess. of Math. 14, 108-112; Coll. Works 12, 300-304.
63. Cayley, A.: On the matricial equation $qQ - Qq' = 0$. Mess. of Math. 14, 176-178; Coll. Works 12, 311-313.
64. Muir, T.: New relations between bipartite functions and determinants with a proof of Cayley's theorem in matrices. Proc. Lond. Math. Soc. 16, 276-286.
65. Sylvester, J. J.: On the trinomial unilateral quadratic equation in matrices of the second order. Quart. Journ. Math. 20, 305-312; Math. Papers 4, 272-277.
66. Weyr, E.: O základní větě v theorii matric, (1884). Sitzb. K. Böhm. Ges. d. Wissens. 1884, 148-152.
67. Weyr, E.: Sur la théorie des matrices. Comptes Rendus 100, 787-789.
68. Weyr, E.: Repartition des matrices en espèces et formation de toutes les espèces. Comptes Rendus 100, 966-969.

1886

69. Bucheim, A.: An extension of a theorem of Professor Sylvester's relating to matrices. Phil. Mag. (5) 22, 173-174.
70. Cayley, A.: On the transformation of the double theta functions. Quart. Journ. Math. 21, 142-178; Coll. Works 12, 358-389.
71. Kumamoto, A.: Matrices no theory ni tsuite shirusu. Tokyo Sugaku Butsurigaku Kwai Kizi, 3, 153-161.

1887

72. Bucheim, A.: On double algebra, (1886). Mess. of Math. 16, 62-63.
73. Bucheim, A.: Note on triple algebra, (1886). Mess. of Math. 16, 111-114.
74. Gibbs, J. W.: Multiple algebra, (1886). Proc. Amer. Assoc. Adv. Sci. 35, 37-66.
75. Peano, G.: Integrazione per serie delle equazioni differenziali lineari. Atti R. Accad. Torino, 22, 437-446. Translated under title "Intégration par séries des équations différentielles linéaires." Math. Ann. 32 (1888), 450-456.

1888

76. Brunel, G.: Sur les racines des matrices zéroïdales. Comptes Rendus 106, 467-470.
77. Bucheim, A.: On a theorem of Prof. Klein's relating to symmetric matrices, (1887). Mess. of Math. 17, 79.
78. Morrice, G.: Multiplication of nonions, (1887). Mess. of Math. 17, 104-105.
79. Weyr, E.: O binarných matricích, (1887). Sitzb. K. Böhm. Ges. d. Wissens. 1887, 358-400.
80. Weyr, E.: Sur la réalisation des systèmes associatifs de quantités complexes à l'aide des matrices, (1887). Sitzb. K. Böhm. Ges. d. Wissens. 1887, 616-618.

1889

81. Bucheim, A.: Note on matrices in involution, (1887). Mess. of Math. 18, 102-104.
82. Sylvester, J. J.: Sur la réduction biorthogonale d'une forme linéolinéaire à sa forme canonique. Comptes Rendus 108, 651-653; Math. Papers 4, 638-640.
83. Weyr, E.: O theorii forem bilinearých. Praze. Translated under title "Zur Theorie der bilinearen Formen" Monatsh. f. Math. u. Phys. 1 (1890), 163-236.

1890

84. Taber, H.: On the theory of matrices. Amer. Journ. Math. 12, 337–396.
85. Voss, A.: Über die conjugirte Transformation einer bilinearen Form in sich selbst, (1889). Münch. Ber. 19, 175–211.
86. Voss, A.: Über die mit einer bilinearen Form vertauschbaren Formen, (1889). Münch. Ber. 19, 283–300.
87. van Wettum, T. B.: De quaternion van Hamilton als matrix van Cayley. Nieuw Archief 17, 206–216.

1891

88. Carvallo, E.: Sur les systèmes linéaires, ... Monatsh. f. Math. u. Phys. 2, 177–216, 225–266, 311–330.
89. Cayley, A.: Note on the involutant of two binary matrices. Mess. of Math. 20, 136–137; Coll. Works 13, 74–75.
90. Cayley, A.: Note on a theorem in matrices. Proc. Lond. Math. Soc. 22, 458; Coll. Works 13, 114.
91. Chapman, C. H. On the matrix which represents a vector. Amer. Journ. Math. 13, 363–380.
92. Kronecker, L.: Reduction der Systeme mit n^2 ganzzahligen Elementen. Crelle 107, 135–136.
93. Molenbroek, P.: Theorie der Quaternionen. Leiden.
94. Rados, G.: Zur Theorie der adjungirte Substitutionen. Math. és Term. Értesitö 1891; Math. Ann. 48 (1897), 417–424.
95. Study, E.: Recurrierende Reihen und bilineare Formen, (1889). Monatsh. f. Math. u. Phys. 2, 23–54.
96. Taber, H.: On certain identities in the theory of matrices, (1890). Amer. Journ. Math. 13, 159–172.
97. Taber, H.: On the application to matrices of any order of the quaternion symbols S and V, (1890). Proc. Lond. Math. Soc. 22, 67–79.
98. Taber, H.: On the matrical equation $\varphi\Omega = \Omega\varphi$. Proc. Amer. Acad. Boston 26, 64–66.
99. Taber, H.: On certain properties of symmetric, skew symmetric, and orthogonal matrices. Proc. Lond. Math. Soc. 22, 449–469.
100. van Wettum, T. B.: Over den quaternion-matrix. Nieuw Archief 18, 168–186.

1892

101. van Elfrinkhof, L.: De oplossing van lineaire vector-verglijkingen in bijzondere gevallen. Nieuw Archief 19, 132–142.
102. van Elfrinkhof, L.: Opmerkingen naar aanleiding der verhandelingen over quaternion-matrices van den heer Th. B. van Wettum. Nieuw Archief 19, 143–150.
103. Metzler, W. H.: On the roots of matrices. Amer. Journ. Math. 14, 326–377.
104. Molien, T.: Über Systeme höherer complexer Zahlen, (1891). Diss. Dorpat; Math. Ann. 41, 83–156.
105. Rados, G.: Zur Theorie der orthogonalen Substitutionen, (1891). Math. és Term. Értesitö 10, 16–18; Math. Natur. Ber. aus Ungarn 10 (1893), 95–97.
106. Rados, G.: Die Theorie der adjungirten Substitutionen, (1891). Math. és Term. Értesitö 10, 34–42; Math. Natur. Ber. aus Ungarn 10 (1893), 98–107.
107. Rost, G.: Untersuchungen über die allgemeinste lineare Substitution, deren Potenzen eine endliche Gruppe bilden. Diss. Würzburg; Leipzig.
108. Voss, A.: Über die cogredienten Transformationen einer bilinearen Form in sich selbst, (1890). Münch. Abh. 17, 235–356.

1893

109. Élie, B.: La fonction vectorielle et ses applications à la physique. Mém. Soc. des Sci. Phys. et Natur. Bordeaux (4) 3, 1–137.
110. McAulay, A.: Utility of quaternions in physics. London.
111. Metzler, W. H.: On certain properties of symmetric, skew symmetric, and orthogonal matrices, (1892). Amer. Journ. Math. 15, 274–282.
112. Taber, H.: On a theorem of Sylvester's relating to non-degenerate matrices. Proc. Amer. Acad. Boston 27, 46–55.
113. Taber, H.: Note on the representation of orthogonal matrices, (1892). Proc. Amer. Acad. Boston 27, 163–164.
114. Taber, H.: On real orthogonal substitution. Proc. Amer. Acad. Boston 28, 212–217.
115. Taber, H.: On the linear transformations between two quadrics. Proc. Lond. Math. Soc. 24, 290–306.

1894

116. Franklin, F.: Note on induced linear substitutions. Amer. Journ. Math. 16, 205–206.
117. Hurwitz, A.: Zur Invariantentheorie. Math. Ann. 45, 381–404.
118. Metzler, W. H.: Matrices which represent vectors. Boston Tech. Quart. 6 (1893), 348–351.
119. Metzler, W. H.: Compound determinants, (1893). Amer. Journ. Math. 16, 131–150.
120. Metzler, W. H.: Homogeneous strains. Annals of Math. 8, 148–156.
121. Peano, G.: Sur les systèmes linéaires. Monatsh. f. Math. u. Phys. 5, 136.
122. Sforza, G.: Sulle forme bilineari simili. Giorn. Mat. 32, 293–316; 33 (1895), 80–105; 34 (1896), 252–278.
123. Taber, H.: On orthogonal substitutions that can be expressed as a function of a single alternate (or skew symmetric) linear substitution, (1893). Amer. Journ. Math. 16, 123–130.
124. Taber, H.: On orthogonal substitutions. Bull. New York Math. Soc. 3, 251–259.
125. Taber, H.: On the automorphic linear transformation of a bilinear form. Proc. Amer. Acad. Boston 29, 178–179.
126. Taber, H.: On the group of automorphic linear transformations of a bilinear form. Proc. Amer. Acad. Boston 29, 371–381.
127. van Wettum, T. B.: Researches on matrices and quaternions. Leyden.

1895

128. Brill, J.: On the application of the theory of matrices to the discussion of linear differential equations with constant coefficients, (1894). Proc. Cambr. Phil. Soc. (3) 8, 201–210.
129. van Elfrinkhof, L.: Der vergelijking $V\rho\varphi\rho = 0$. Nieuw Archief (2) 1, 76–87.
130. van Elfrinkhof, L.: Draaiings-matrices en quaternions. Nieuw Archief (2) 1, 88–100.
131. Hensel, K.: Über die Elementartheiler componirter Systeme, (1894). Crelle 114, 109–115.
132. Joly, C. J.: The theory of linear vector functions, (1894). Trans. Roy. Irish Acad. 30, pt. 16, 597–647.
133. Schlesinger, L.: Handbuch der Theorie der linearen Differentialgleichungen, p. 91 ff. Leipzig.
134. Taber, H.: On the automorphic linear transformation of an alternate bilinear form. Math. Ann. 46, 561–583.
135. Taber, H.: On those orthogonal substitutions that can be generated by the repetition of an infinitesimal orthogonal substitution. Proc. Lond. Math. Soc. 26, 364–376.
136. van Wettum, T. B.: Over het quotient van twee ruimte-vectoren, en over een quaternion. Nieuw Archief (2) 1, 68–75.

1896

137. BRILL, J.: Note on matrices, (1895). Proc. Lond. Math. Soc. 27, 35–38.
138. FROBENIUS, G.: Zur Theorie der Scharen bilinearer Formen, (1881). Zürich Naturf. Ges. 41, T. 2, 20–23.
139. FROBENIUS, G.: Über die cogredienten Transformation der bilinearen Formen. Berlin Sitzb. 1896, 7–16.
140. FROBENIUS, G.: Über vertauschbare Matrizen. Berlin Sitzb. 1896, 601–614.
141. JOLY, C. J.: Scalar invariants of two linear vector functions, (1895). Trans. Roy. Irish Acad. 30 pt. 18, 709–728.
142. JOLY, C. J.: Quaternion invariants of linear vector functions and quaternion determinants. Proc. Roy. Irish Acad. (3) 4 pt. 1, 1–15.
143. LANDSBERG, G.: Über Fundamentalsysteme und bilineare Formen, (1895). Crelle 116, 331–349.
144. LAURENT, H.: Exposé d'une théorie nouvelle des substitutions linéaires. Nouv. Ann. (3) 15, 345–365.
145. PINCHERLE, S.: Le operazioni distributive e le omografie. Rend. R. Ist. Lomb. (2) 29, 397–405.
146. RADOS, G.: Adjungirte quadratische Formen. Math. és Term. Értesitö 14, 26–32; Math. Natur. Ber. aus Ungarn 14 (1898), 85–91.
147. RADOS, G.: Zur Theorie der adjungirten bilinearen Formen. Math. és Term. Értesitö 14, 165–175; Math. Natur. Ber. aus Ungarn 14 (1898), 116–127.
148. TABER, H.: On orthogonal substitution, (1893). Math. Papers, Intern. Congr. Chicago, 395–400.
149. TABER, H.: Note on the automorphic linear transformation of a bilinear form, (1895). Proc. Amer. Acad. Boston 31, 181–192.

1897

150. BAKER, H. F.: Abelian functions, App. II. Cambridge.
151. BRILL, J.: Supplementary note on matrices, (1896). Proc. Lond. Math. Soc. 28, 368–370.
152. LAURENT, H.: Applications de la théorie des substitutions linéaires à l'étude des groupes. Nouv. Ann. (3) 16, 149–168.
153. LOEWY, A.: Bemerkung zur Theorie der konjugirten Transformation einer bilinearen Form in sich selbst, (1896). Münch. Ber. 26, 25–30.
154. LOEWY, A.: Zur Theorie der linearen Substitutionen. Math. Ann. 48, 97–110.
155. SHAW, J. B.: The linear vector operator of quaternions, (1895). Amer. Journ. Math. 19, 267–282.
156. TABER, H.: On the transformations between two symmetric or alternate bilinear forms. Math. Review 1, 120–126.
157. TABER, H.: On the group of real linear transformations whose invariant is a real quadratic form, (1896). Math. Review 1, 154–168; Proc. Amer. Acad. Boston 32, 77–83.
158. TABER, H.: Notes on the theory of bilinear forms. Bull. Amer. Math. Soc. 3, 156–164.
159. TAIT, P. G.: On the linear and vector function, (1896). Proc. Roy. Soc. Edinb. 21, 160–164; Sci. Papers 2, 406–409.
160. TAIT, P. G.: On the linear and vector function. Proc. Roy. Soc. Edinb. 21, 310–312; Sci. Papers 2, 410–412.
161. TAIT, P. G.: Note on the solution of equations in linear and vector functions. Proc. Roy. Soc. Edinb. 21, 497–505; Sci. Papers 2, 413–420.
162. VOSS, A.: Über die Anzahl der cogredienten und adjungirten Transformationen einer bilinearen Form in sich selbst, (1896). Münch. Ber. 26, 211–272.
163. VOSS, A.: Symmetrische und alternirende Lösungen der Gleichung $SX = XS'$, (1896). Münch. Ber. 26, 273–281.

1898

164. Bachmann, P.: Zahlentheorie, vol. 4, p. 275 ff. Leipzig.
165. Joly, C. J.: The associative algebra applicable to hyperspace, (1897). Proc. Roy. Irish Acad. (3) 5, 73–123.
166. Laurent, H.: Exposé d'une théorie nouvelle des substitutions. Journ. de Math. (5) 4, 75–119.
167. Loewy, A.: Über die bilineare Formen mit conjugirt imaginären Variablen, (1897). Nova Acta Leop-Carol. Acad. 71, 379–446.
168. Loewy, A.: Über bilineare Formen mit conjugirt imaginären Variablen. Math. Ann. 50, 557–576.
169. Muir, T.: A reinvestigation of the problem of the automorphic linear transformation of a bipartite quadric. Amer. Jour. Math. 20, 215–228.
170. Rados, G.: Zur Theorie der adjungirten quadratischen Formen, (1897). Verh. Intern. Math. Congr. Zürich 1, 163–165.
171. Ravut, L.: Remarques sur une matrice. Nouv. Ann. (3) 17, 118–120.
172. Stephanos, C.: Sur un mode de composition des déterminants et des formes bilinéaires. Giorn. Mat. 36, 376–379.
173. Whitehead, A. N.: Universal algebra, vol. 1, pp. 248–269, 318–346. Cambridge.

1899

174. Baker, H. F.: (Note on a paper of Prof. Burnside's). Proc. Lond. Math. Soc. 30, 195–198.
175. Muth, P.: Theorie und Anwendung der Elementartheiler. Leipzig.
176. Rados, G.: Gruppen inducierter Substitutionen, (1898). Math. és Term. Értesitö 17, 44–65; Math. Natur. Ber. aus Ungarn 17 (1901), 227–247.
177. Shaw, J. B.: Some generalisations in multiple algebra and matrices. Bull. Amer. Math. Soc. 5, 381–382.
178. Stephanos, C.: Sur une extension du calcul des substitutions linéaires. Comptes Rendus 128, 593–596.
179. Study, E.: Theorie der gemeinen und höheren complexen Grössen. Encycl. der Math. Wiss. I 1, 413 ff.

1900

180. Amaldi, U.: Sulle sostituzioni lineari commutabili. Rend. R. Ist. Lomb. (2) 33, 731–744.
181. Bendixson, J.: Sur les racines d'une équation fondamentale. Öfv. K. Svenska Vet.-Akad. Förh. Stockholm 57, 1099–1103.
182. Loewy, A.: Über Scharen reeller quadratischer und Hermitescher Formen. Crelle 122, 53–72.
183. Loewy, A.: Über die Transformation einer Hermiteschen Form von nicht verschwindender Determinante in sich. Gött. Nachr. 1900, 298–302.
184. Schlesinger, L.: Über vertauschbare lineare Substitutionen. Crelle 121, 177–187.
185. Stephanos, C.: Sur une extension du calcul des substitutions linéaires. Journ. de Math. (5) 6, 73–128.
186. Tait, P. G.: On the linear and vector function, (1899). Proc. Roy. Soc. Edinb. 22, 547–549; Sci. Papers 2, 424–426.

1901

187. Bromwich, T. J. I'A.: Canonical reduction of linear substitutions and bilinear forms with a dynamical application, (1900). Proc. Lond. Math. Soc. 32, 79–118, 321–352.
188. Bromwich, T. J. I'A.: Note on Weierstrass's reduction of a family of bilinear forms, (1900). Proc. Lond. Math. Soc. 32, 158–163.
189. Bromwich, T. J. I'A.: On a canonical reduction of bilinear forms with special consideration of congruent reductions, (1900). Proc. Lond. Math. Soc. 32, 321–352.

190. Bromwich, T. J. I'A.: Theorems on matrices and bilinear forms, (1900). Proc. Cambr. Phil. Soc. 11, 75–89.

191. Bromwich, T. J. I'A.: Ignoration of coordinates as a problem in linear substitutions. Proc. Cambr. Phil. Soc. 11, 163–167.

192. Bucheim, A.: Review: "Theorie und Anwendung der Elementartheiler" by Dr. P. Muth. Bull. Amer. Math. Soc. 7, 308–316.

193. Carlini, L.: Sul prodotto di due matrici rettangolari coniugate, (1900). Period. di Mat. (2) 3, 193–198.

194. Gibbs, J. W., and E. B. Wilson: Vector analysis, chap. 5. (Dover Reprint)

195. Joly, C. J.: Hamilton's "Elements of quaternions," 2nd ed., appendix. London.

196. Loewy, A.: Über die Verallgemeinerung eines Weierstrassschen Satzes. Crelle 123, 258–262.

197. Plemelj, J.: Ein Satz über vertauschbare Matricen und seine Anwendung in der Theorie linearer Differentialgleichungen. Monatsh. f. Math. u. Phys. 12, 82–96.

198. Schur, I.: Über eine Klasse von Matrizen die sich einer gegebenen Matrix zuordnen lassen. Diss. Berlin.

1902

199. Autonne, L.: Sur les groupes linéaires, réels, et orthogonaux. Bull. Soc. Math. France 30, 121–134.

200. Autonne, L.: Sur l'Hermitien, (1901). Rend. Circ. Mat. Palermo 16, 104–128.

201. Bendixson, J.: Sur les racines d'une équation fondamentale. Acta Math. 25, 359–365.

202. Burnside, W.: On the characteristic equations of certain linear substitutions. Quart. Journ. Math. 33, 80–84.

203. Carlini, L.: Sopra due tipi di relazioni . . . , (1901). Period. di Mat. (2) 4, 175–179.

204. Dickson, L. E.: A matrix defined by the quaternion group. Amer. Math. Monthly 9, 243–248.

205. Hensel, K., and G. Landsberg: Theorie der algebraischen Funktionen einer Variabeln, p. 174 ff. Leipzig.

206. Hirsch, A.: Sur les racines d'une équation fondamentale. Acta Math. 25, 367–370.

207. Niccolletti, O.: Sulle matrici associate ad una matrice data. Atti R. Accad. Torino 37, 655–659.

208. Schur, I.: Über einen Satz aus der Theorie der vertauschbaren Matrizen. Berlin Sitzb. 1902, 120–125.

209. Volterra, V.: Sui fondamenti della teoria delle equazioni differenziali lineari. Mem. Soc. Ital. Sci. (3) 12, 3–68.

1903

210. Autonne, L.: Sur la décomposition d'une substitution linéaire, réelle et orthogonale, en un produit d'inversions. Comptes Rendus 136, 1185–1186.

211. Autonne, L.: Sur la canonisation des formes bilinéaires. Nouv. Ann. (4) 3, 57–64.

212. Autonne, L.: Sur l'hypohermitien. Bull. Soc. Math. France 31, 140–155.

213. Autonne, L.: Sur quelques propriétés des matrices hypohermitiennes. Bull. Soc. Math. France 31, 268–271.

214. Autonne, L.: Sur la décomposition d'une substitution linéaire, réelle et orthogonale, en un produit d'inversions. Ann. Univ. Lyon I, fasc. 12.

215. Baker, H. F.: On some cases of matrices with linear invariant factors, (1902). Proc. Lond. Math. Soc. 35, 379–384.

216. Baker, H. F.: On the invariant factors of a determinant. Proc. Cambr. Phil. Soc. 12, 65–77.

217. Frobenius, G.: Theorie der hypercomplexen Grössen. Berlin Sitzb. 1903, 504–537, 634–645.

218. Hudson, R. W. H. T.: Matrix notation in the theory of screws, (1902). Mess. of Math. 32, 51–57.
219. Joly, C. J.: The multilinear quaternion function, (1902). Proc. Roy. Irish Acad. 24, 47–52.
220. Joly, C. J.: Quaternions and projective geometry. Trans. Roy. Soc. Lond. 201, 223–327.
221. Rados, G.: Notes sur les substitutions orthogonales. Math. Natur. Ber. aus Ungarn 18, 231–235.
222. Rados, G.: Zur Theorie der algebraischen Resolventen. Math. Natur. Ber. aus Ungarn 18, 236–249.
223. Shaw, J. B.: Theory of linear associative algebra, (1902). Trans. Amer. Math. Soc. 4, 251–287.
224. Wedderburn, J. H. M.: On the applications of quaternions in the theory of differential equations. Trans. Roy. Soc. Edinb. 40, 709–721.
225. Wellstein, J.: Über die Frobeniusschen Kovarienten einer Bilinearform, (1901). Arch. der Math. u. Phys. (3) 5, 229–241.

1904

226. Baker, H. F.: Note on Sylvester's theorems on determinants. Collected Math. Papers of J. J. Sylvester, 1, 647–650.
227. Burnside, W.: On linear substitutions of determinant unity with integral coefficients. Mess. of Math. 33, 133–137.
228. Hensel, K.: Theorie der Körper von Matrizen. Crelle 127, 116–166.
229. Laurent, H.: Sur les transformations qui transforment une forme du second degré donnée en une autre également donnée. Nouv. Ann. (4) 4, 29–38.
230. de Séguier, J. A.: Éléments de la théorie des groupes abstraits, p. 159 ff. Paris.
231. Taber, H.: On the real automorphic linear transformation of a real bilinear form, (1903). Proc. Amer. Acad. Boston 39, 307–320.
232. Wedderburn, J. H. M.: Note on the linear matrix equation. Proc. Edinb. Math. Soc. 22, 49–53.

1905

233. Bromwich, T. J. I'A.: On the roots of the characteristic equation of a linear substitution, (1904). Brit. Assoc. Report 1904, 440–441.
234. Cunningham, E.: On the normal series satisfying linear differential equations, (1904). Phil. Trans. Roy. Soc. Lond. 205, 1–35.
235. Joly, C. J.: A manual of quaternions, chap. 8. London.
236. Muth, P.: Über reelle Äquivalenz von Scharen reeller quadratischer Formen. Crelle 128, 302–321.
237. Schlesinger, L.: Beiträge zur Theorie der Systeme linearer homogener Differentialgleichungen. Crelle 128, 263–297.
238. Schur, I.: Zur Theorie der vertauschbaren Matrizen. Crelle 130, 66–76.

1906

239. Autonne, L.: Sur les polynomes à coefficients et à variable hypercomplexes. Bull. Soc. Math. France 34, 205–212.
240. Bromwich, T. J. I'A.: On the roots of the characteristic equation of a linear substitution, (1904). Acta Math. 30, 297–304.
241. Gibbs, J. W.: Elements of vector analysis, chap. 3, (1884). Sci. Papers 2, 53 ff.
242. Kreis, H.: Contribution à la théorie des systémes linéaires. Thése, Zurich.
243. Petr, K.: Několik poznámek o determinantech. Casopis pro Pest. Math. u. Fys. 35, 311–321.

244. Petr, K.: Néhány megjegyzés a determinánsok elmeletéhez. Math. es Phys. Lapok 15, 353–365; translated under title 'Einige Bemerkungen über die Determinanten.' Math. Natur. Ber. aus Ungarn 25 (1908), 95–105.

245. Wedderburn, J. H. M.: On a theorem in hypercomplex numbers. Proc. Roy. Soc. Edinb. 26, 48–50.

1907

246. Wirth, J.: Über die Elementarteiler einer linearen homogenen Substitution. Diss. Freiburg.

247. Bôcher, M.: Introduction to higher algebra, chap. 8–13, 20–22. New York.

248. Brill, J.: On the expression of the so-called biquaternions and triquaternions with the aid of quaternary matrices, (1906). Proc. Lond. Math. Soc. (2) 4, 124–130.

249. Perron, O.: Zur Theorie der Matrices, (1906). Math. Ann. 64, 248–263.

250. de Séguier, J.-A.: Sur la théorie des matrices. Comptes Rendus 145, 1259–1260.

251. Shaw, J. B.: Synopsis of linear associative algebras. Washington, D. C.

252. Wedderburn, J. H. M.: On hypercomplex numbers. Proc. Lond. Math. Soc. (2) 6, 77–118.

1908

253. Cartan, E. and E. Study.: Nombres complexes. Encycl. des Sci. Math. I 1, fasc. 3, 413 ff.

254. Frobenius, G.: Über Matrizen aus positiven Elementen. Berlin Sitzb. 1908, 471–476.

255. Kreis, H.: Auflösung der Gleichung $X^n = A$. Zürich Naturf. Ges. 53, 366–376.

256. Nicoletti, O.: Sulla riduzione a forma canonica di una sostituzione lineare omogenea e di un fascio di forme bilineari. Ann. di Mat. (3) 14, 265–326.

257. Ranum, A.: Concerning linear substitutions of finite period with rational coefficients, (1907). Trans. Amer. Math. Soc. 9, 183–202.

258. Schlesinger, L.: Vorlesungen über lineare Differentialgleichungen. Leipzig.

259. de Séguier, J.: Sur la théorie des matrices. Bull. Soc. Math. France 36, 20–40.

260. de Séguier, J.: Sur les formes bilinéaires. Comptes Rendus 146, 1247–1248.

261. Wilson, E. B.: On the theory of double products and strains in hyperspace, (1907). Connecticut Acad. Trans. 14, 1–57.

1909

262. Autonne, L.: Sur les groupes de matrices linéaires non invertibles. Ann. Univ. Lyon I fasc. 25.

263. Ball, R.: Linear vector functions, (1908). Brit. Assoc. Report 1908, 611–612.

264. Cecioni, F.: Sulle equazioni fra matrici $AX = XB$, $X^m = A$. R. Accad. dei Lincei Rend. (5) 18, 566–571.

265. Cecioni, F.: Sulla caratteristica del prodotto di due matrici. Period. di Mat. (3) 6, 253–265.

266. Dickson, L. E.: On commutative linear groups, (1907). Quart. Journ. Math. 40, 167–196.

267. Frobenius, G.: Matrizen aus positiven Elementen. II. Berlin Sitzb. 1909, 514–518.

268. Gamberini, G.: Una speciale classe di matrici quadrate permutabili, (1908). Giorn. di Mat. 47, 137–155.

269. Gamberini, G.: Alcuni risultati intorno al prodotto di due matrici quadrate d'ordine n. Rome.

270. Gamberini, G.: Sulle relazioni fra alcune forme del prodotto di due matrici quadrate d'ordine n. Rome.

271. Jackson, D.: Resolution into involutory substitutions of the transformations of a non-singular bilinear form into itself. Trans. Amer. Math. Soc. 10, 479–484.

272. Karst, L.: Lineare Funktionen und Gleichungen. Prog. (127) Realprogymn. Lichtenberg bei Berlin.
273. Schur, I.: Über die characteristischen Wurzeln einer linearen Substitution mit einer Anwendung auf die Theorie der Integralgleichungen. Math. Ann. 66, 488–510.
274. de Séguier, J. A.: Sur les formes bilinéaires et quadratiques. Journ. de Math. (6) 5, 1–63.
275. Wellstein, J.: Die Dekomposition der Matrizen. Gött. Nachr. 1909, 77–99.
276. Wellstein, J.: Kriterien für die Potenzen einer Determinante. Math. Ann. 67, 490–497.
277. Widder, W.: Untersuchungen über die allgemeinste lineare Substitution mit vorgegebener p^{ter} Potenz, (1908). Diss. Würzburg; Leipzig.

1910

278. Autonne, L.: Sur les matrices linéaires échangeables à une matrice donnée. Jour. Ec. Poly. (2) 14, 83–131.
279. Cecioni, F.: Sopra alcune operazioni algebriche sulle matrici, (1909). Ann. R. Scuola Norm. Sup. Pisa 11.
280. Frobenius, G.: Über die mit einer Matrix vertauschbaren Matrizen. Berlin Sitzb. 1910, 3–15.
281. Hawkes, H. E.: The reduction of families of bilinear forms, (1908). Amer. Journ. Math. 32, 101–114.
282. Hutchinson, J. I.: On linear transformations which leave an Hermitian form invariant, (1909). Amer. Journ. Math. 32, 195–206.
283. Kreis, H.: Einige Anwendungen der Matricestheorie. Progr.Gymn. Winterthur.
284. Loewy, A.: Pascal's "Repertorium der höheren Analysis." Leipzig, 2 ed., chap. 2, 79–153.

1911

285. Burgess, H. T.: The simultaneous reduction of a quadratic and a bilinear form by the same transformation on both x's and y's, (1910; abstract). Bull. Amer. Math. Soc. 17, 65–66.
286. Burgess, H. T.: The application of matrices to cubic forms. (Abstract.) Bull. Amer. Math. Soc. 17, 299.
287. Châtelet, A.: Sur certains ensembles de tableaux et leur application à la théorie des nombres. Ann. Ec. Norm. (3) 28, 105–202.
288. David, L.: (On matrices of algebraic iteration; in Magyar.) Math. és Term. Értesitö 29, 444–449.
289. Elie, B.: Relations entre les paramètres Cayléens de trois substitutions orthogonales dont l'une est égale au produit des deux autres. Bull. des Sci. Math. (2) 35, 162–180.
290. Frobenius, G.: Über den Rang einer Matrix. Berlin Sitzb. 1911, 20–29, 128–129.
291. Frobenius, G.: Über den von L. Bieberbach gefundenen Beweis eines Satzes von C. Jordan. Berlin Sitzb. 1911, 241–248.
292. Frobenius, G.: Über unitäre Matrizen. Berlin Sitzb. 1911, 373–378.
293. Rabinovic, J.: (Linear vector functions; in Russian.) Odessa.
294. Ranum, A.: The general term of a recurring series. Bull. Amer. Math. Soc. 17, 457–461.

1912

295. Amaldi, U.: Sulle sostituzioni lineari commutabili. Rend. R. Ist. Lomb. (2) 45, 433–445.
296. Autonne, L.: Sur une propriété des matrices linéaires. Nouv. Ann. (4) 12, 118–127.

297. DRACH, J., AND W. F. MEYER: Théorie des formes et des invariants. Encycl. des Sci. Math. I 2, fasc. 4, 440 ff.
298. FROBENIUS, G.: Über Matrizen aus nicht negativen Elementen. Berlin. Sitzb. 1912, 456–477.
299. HILTON, H.: On properties of certain linear homogeneous substitutions, (1911). Proc. Lond. Math. Soc. (2) 10, 273–283.
300. HILTON, H.: On cyclant substitutions, (1911). Mess. of Math. 41, 49–51.
301. HILTON, H.: Substitutions permutable with a canonical substitution, (1911). Mess. of Math. 41, 110–118.
302. HILTON, H.: On symmetric and orthogonal substitutions. Mess. of Math. 41, 146–154.
303. LATTÈS, S.: Sur la réduction des substitutions linéaires. Comptes Rendus 155, 1482–1484.

1913

304. AUTONNE, L.: Sur les matrices hypohermitiennes et les unitaires. Comptes Rendus 156, 858–860.
305. BIRKHOFF, G. D.: A theorem on matrices of analytic functions. Math. Ann. 74, 122–133.
306. CULLIS, C. E.: Matrices and determinoids, vol. 1. Cambridge.
307. HILTON, H.: Some properties of symmetric and orthogonal substitutions, (1912). Proc. Lond. Math. Soc. (2) 12, 94–99.
308. LOEWY, A.: Über lineare homogene Differentialsysteme und ihre Sequenten. Sitzb. Heidl. Akad. 1913, A17.
309. WEDDERBURN, J. H. M.: Note on the rank of a symmetrical matrix. Annals of Math. (2) 15, 29.

1914

310. BLISS, G. A.: A note on symmetric matrices. Annals of Math. (2) 15, 43–44.
311. CIPOLLA, M.: Le sostituzioni ortogonali non-cayleyane, (1913). Atti Accad. Gioenia, Catania (5) 7, mem. 2, 1–18.
312. DICKSON, L. E.: Linear algebras, chap. 1. Cambridge.
313. HILTON, H.: Properties of certain homogeneous linear substitutions. Annals of Math. (2) 15, 195–201.
314. HILTON, H.: Homogeneous linear substitutions. Oxford.
315. METZLER, W. H.: On the rank of a matrix, (1913). Annals of Math. (2) 15, 161–165.
316. STEIN, J.: Beiträge zur Matrizenrechnung mit Anwendung auf die Relativitätstheorie. Tübingen.
317. WEDDERBURN, J. H. M.: On continued fractions in non-commutative quantities. Annals of Math. (2) 15, 101–105.
318. WEDDERBURN, J. H. M.: Note on the rank of a symmetrical matrix. II. Annals of Math. (2) 16, 86–88.

1915

319. AMATO, V.: Sulla forma canonica delle sostituzioni ortogonali periodiche. Atti Accad. Gioenia, Catania (5) 8, mem. 15, 1–8.
320. AUTONNE, L.: Sur les matrices hypohermitiennes et sur les matrices unitaires. Ann. Univ. Lyon I fasc. 38.
321. BOEHM, K.: Über einen Determinantensatz, in welchem das Multiplikationstheorem als besonderer Fall enthalten ist. Crelle 145, 250–253.
322. KRAWCZUK, M.: (On groups of permutable matrices; in Russian.) Soc. Math. Kharkoff (2) 14, 163–176.
323. LATTÈS, S.: Sur les multiplicités linéaires invariantes par une substitution linéaire donnee. Comptes Rendus 160, 671–674.

324. PERRON, O.: Über konvergente Matrixprodukte. Sitzb. Heidl.Akad. 6, A4, 1–27.
325. WEDDERBURN, J. H. M.: On matrices whose coefficients are functions of a single variable. Trans. Amer. Math. Soc. 16, 328–332.

1916

326. BIRKHOFF, G. D.: Infinite products of analytic matrices. Trans. Amer. Math. Soc. 17, 386–404.
327. BURGESS, H. T.: A practical method of determining elementary divisors. Annals of Math. 18, 4–6.
328. BURGESS, H. T.: On the matrix equation $BX = C$. Amer. Math. Monthly 23, 152–155.
329. CULLIS, C. E.: Primitive matrices and the primitive degrees of a matrix. Bull. Calcutta Math. Soc. 6, 35–54.
330. LATTÈS, S.: Sur une forme canonique des substitutions linéaires. Ann. Toulouse (3) 6, 1–84.
331. LEVI, B.: Introduzione alla analisi matematica. Paris.
332. SCORZA, G.: Intorno alla teoria generale delle matrici di Riemann e ad alcune sue applicazioni. Rend. Circ. Mat. Palermo 41, 263–379.
333. TABER, H.: Conditions for the complete reducibility of groups of linear substitutions. Amer. Journ. Math. 38, 337–372.
334. WHITTAKER, E. T.: On the theory of continued fractions. Proc. Roy. Soc. Edinb. 36, 243–255.
335. WIDDER, W.: Über orthogonale, involutorische und orthogonal-involutorische Substitutionen. Progr. K. Alten Gymn. Würzburg.

1917

336. CULLIS, C. E.: Primitive matrices and the primitive degrees of a matrix. Pt. II. Bull. Calcutta Math. Soc. 8, 1–32.
337. KOWALEWSKI, G.: Natürliche Normalform linearer Transformationen. Leipz. Ber. 69, 325–335.
338. LOEWY, A.: Die Begleitmatrix eines linearen homogenen Differentialausdruckes. Gött. Nachr. 1917, 255–263.
339. SZÁSZ, O.: Über eine Verallgemeinerung des Hadamardschen Determinantensatzes. Monatsh. f. Math. u. Phys. 28, 253–257.

1918

340. BURGESS, H. T.: Solution of the matrix equation $X^{-1}AX = N$. Annals of Math. 19, 30–36.
341. CULLIS, C. E.: Matrices and determinoids, vol. 2. Cambridge.
342. LOEWY, A.: Über Matrizen- und Differentialkomplexe, (1915–17). Math. Ann. 78, 1–51, 343–358, 359–368.
343. LOEWY, A.: Über einen Fundamentalsatz für Matrizen oder lineare homogene Differentialsysteme. Sitzb. Heidl. Akad. 1918, A5, 1–36.
344. MOORE, C. L. E., AND H. B. PHILLIPS: The dyadics which occur in a point-space of three dimensions. Proc. Amer. Acad. Boston 53, 387–438.
345. MUIR, T.: Note on the construction of an orthogonant. Proc. Roy. Soc. Edinb. 38, 146–153.
346. SCORZA, G.: Sopra alcune notevoli matrici riemanniane. Atti R. Accad. Sci. Torino 53, 598–607.
347. STUDY, E.: Zur Theorie der linearen Gleichungen. Acta Math. 42, 1–61.
348. TOEPLITZ, O.: Das algebraische Analogen zu einen Satz von Fejer. Math. Zeits. 2, 187–197.

1919

349. Bennett, A. A.: Products of skew symmetric matrices. Bull. Amer. Math. Soc. 25, 455–458.
350. Phillips, H. B.: Functions of matrices. Amer. Journ. Math. 41, 266–278.
351. Speiser, A.: Zahlentheoretische Sätze aus der Gruppentheorie. Math. Zeit. 5, 1–6.

1920

352. Bose, A. C.: Review: Cullis's Matrices and Determinoids. Bull. Calcutta Math. Soc. 10, 243–256; 11, 51–82.
353. Coolidge, J. L.: The geometry of Hermitian forms, (1918). Trans. Amer. Math. Soc. 21, 44–51.
354. Ghosh, N.: Potent divisors of the characteristic matrix of a minimum simple square anti-slope. Bull. Calcutta Math. Soc. 11, 1–6.
355. Leveugle, R.: Précis de calcul géométrique, chap. 9. Paris.
356. Loewy, A.: Begleitmatrizen und lineare homogene Differentialausdrücke, (1919). Math. Zeitschr. 7, 58–125.
357. Pidduck, F. B.: Functions of limiting matrices. Proc. Lond. Math. Soc. (2) 19, 398–408.

1921

358. Cullis, C. E.: Evaluation of the product matrix in a commutantal product of simple matrices having given nullities, (1920). Bull. Calcutta Math. Soc. 11, 105–150.
359. Krull, W.: Über Begleitmatrizen und Elementarteilertheorie. Diss. Freiburg.
360. Scorza, G.: Alcune proprietà delle algebre regolari. Note e Memorie di Mat. 1, 198–209.
361. Scorza, G.: Le algebre di ordine qualunque e le matrici di Riemann. Rend. Circ. Mat. Palermo 45, 1–204.
362. Scorza, G.: Corpi numerici e algebre. Messina.
363. Szász, O.: Über Hermitesche Formen mit rekurrierender Determinante und über rationale Polynome, (1920). Math. Zeits. 11, 24–57.

1922

364. Dahr, S.: On the inverse of an undegenerate non-plural quadrate slope, (1921). Bull. Calcutta Math. Soc. 12, 85–92.
365. Hitchcock, F. L.: A solution of the linear matrix equation by double multiplication. Proc. Amer. Nat. Acad. Sci. 8, 78–83.
366. Logsdon, M. I.: Equivalence and reduction of pairs of Hermitian forms, (1921). Amer. Journ. Math. 44, 247–260.
367. Radon, J.: Lineare Scharen orthogonaler Matrizen, (1921). Hamburg Abh. 1, 1–14.
368. Schur, I.: Über Ringbereiche im Gebiete der ganzzahligen linearen Substitutionen. Berlin Sitzb. 1922, 145–168.
369. Shaw, J. B.: Vector calculus, chap. 9. New York.
370. Veblen, O., and P. Franklin. On matrices whose elements are integers, (1921). Annals of Math. (2) 23, 1–15.
371. Wedderburn, J. H. M.: The automorphic transformation of a bilinear form, (1921). Annals of Math. 23, 122–134.

1923

372. Birkhoff, G. D., and R. E. Langer.: The boundary problems and developments associated with a system of ordinary linear differential equations of the first order. Proc. Amer. Acad. Boston 58, 51–128.
373. Brahana, H. R.: A theorem concerning certain unit matrices with integer elements. Annals of Math. 24, 265–270.

374. DICKSON, L. E.: Algebras and their arithmetics. Chicago.
375. HASSE, H.: Symmetrische Matrizen im Körper der rationalen Zahlen. Crelle 153, 12–43.
376. HITCHCOCK, F. L.: On double polyadics, with application to the linear matrix equation. Proc. Amer. Acad. Boston, 58, 353–395.
377. MEHMKE, R.: Einige Sätze über Matrizen, (1922). Crelle 152, 33–39.
378. WEINSTEIN, A.: Fundamentalsatz der Tensorrechnung, (1922). Math. Zeitschr. 16, 78–91.
379. WELTZIEN, C.: Die n-te Wurzel aus einer homogenen linearen Substitution von 3 Veränderlichen. Sitzb. Berl. Math. Ges. 22, 48.

1924

380. DICKSON, L. E.: Algebras and their arithmetics. Bull. Amer. Math. Soc. 30, 247–257.
381. KRAVČUK, M. (Quadratic forms and linear transformations; in Ukrainian.) Mem. Kiev Acad. Sci. 1_3.
382. KRAVČUK, M.: (On the theory of commutative matrices; in Ukrainian.) Bull. Kiev Acad. Sci. 1, 28–33.
383. KÜRSCHÁK, J.: On matrices connected with Sylvester's dialytic eliminant. Trans. Roy. Soc. So. Africa 11, 257–261.
384. MUIR, T.: Note on a property of bigradient arrays connected with Sylvester's dialytic eliminant. Trans. Roy. Soc. So. Africa, 11, 101–104.

1925

385. BORN, M., AND P. JORDAN. Zur Quantenmechanik. Zeitschr. f. Phys. 34, 858–888.
386. CULLIS, C. E.: Matrices and determinoids, vol. 3, pt. 1. Cambridge.
387. KÜRSCHÁK, J.: Speciális Mátrixokról. Math. és Phys. Lapok 32, 9–13.
388. SCHUR, I.: Einige Bemerkungen zur Determinantentheorie. Sitzb. Akad. Berlin 1925, 454–463.
389. WEDDERBURN, J. H. M.: The absolute value of the product of two matrices. Bull. Amer. Math. Soc. 31, 304–308.

1926

390. BECK, H.: Einführung in die Axiomatik der Algebra. Berlin.
391. BORN, M., W. HEISENBERG AND P. JORDAN: Zur Quantenmechanik. II, (1925). Zeitschr. f. Phys. 53, 557–615.
392. DICKSON, L. E.: Modern algebraic theories. (Dover Reprint)
393. DUBNOV, J. S.: (Redoubled symmetric orthogonal matrices; in Russian.) Moscow State Univ. Ins. Math. Mech. 1926, 33–54.
394. HENSEL, K.: Über Potenzenreihen von Matrizen. Crelle 155, 107–110.
395. KAGAN, V. F.: (On some number-systems following from the Lorentz transformation; in Russian.) Moscow State Univ. Inst. Math. Mech. 1926, 3–31.
396. KRULL, W.: Theorie und Anwendung der verallgemeinerten Abelschen Gruppen. Sitzb. Heidl. Akad. 1926, 1.
397. MACDUFFEE, C. C.: The nullity of a matrix relative to a field, (1925). Annals of Math. 27, 133–139.
398. RILEY, J. L.: Theory of matrices over any division algebra. Bull. Calcutta Math. Soc. 17, 1–12.
399. WEDDERBURN, J. H. M.: Note on matrices in a given field. Annals of Math. 27, 245–248.

1927

400. ARAMATA, K.: Über einen Satz für unitäre Matrizen. Tôhoku Math. Journ. 28, 281.
401. DICKSON, L. E.: Algebren und ihre Zahlentheorie. Zürich.

402. Fischer, E.: Über die Cayleysche Eliminationsmethode, (1926). Math. Zeitschr. 26, 497–550.

403. Frisch, R.: Sur le théorème des déterminants de M. Hadamard. Comptes Rendus 185, 1244–1245.

404. Krawtchouk, M.: Über vertauschbare Matrizen. Rend. Circ. Mat. Palermo 51, 126–130.

405. von Neumann, J.: Zur Theorie der Darstellungen kontinuierlicher Gruppen. Berlin Sitzb. 1927, 76–90.

406. Porcu-Tortrini, E.: Sulle potenze delle matrici di secondo ordine. Atti Pontificia Accad. 80, 150–153.

407. Porcu-Tortrini, E.: Calcolo delle potenze delle matrici di secondo ordine mediante riduzione alla forma canonica. Atti Pontificia Accad. 80, 277–281.

408. Porcu-Tortrini, E.: Terzo procedimento del calcolo delle potenze delle matrici di second' ordine. Atti Pontificia Accad. 80, 348–353.

409. Schur, I : Über die rationalen Darstellungen der allgemeinen linearen Gruppe. Berlin Sitzb. 1927, 58–75.

410. Spampinato, N.: Nuovi contributi alla teoria generale delle matrici di Riemann. Rend. Circ. Mat. Palermo 51, 238–258.

411. Turnbull, H. W.: The matrix square and cube roots of unity. Journ. Lond. Math. Soc. 2, 242–244.

1928

412. Aitken, A. C.: On the latent roots of certain matrices, (1927). Proc. Edinb. Math. Soc. (2) 1, 135–138.

413. Aitken, A. C.: Note on the elementary divisors of some related matrices. Proc. Edinb. Math. Soc. (2) 1, 166–168.

414. Brauer, R.: Über einen Satz für unitäre Matrizen. Tôhoku Math. Journ. 30, 72.

415. Browne, E. T.: Involutions that belong to a linear class. Annals of Math. 29, 483–492.

416. Burgatti, P.: Sulle equazioni algebriche a matrice. Boll. Un. Mat. Ital. 7, 65–69.

417. Dickson, L. E.: A new theory of linear transformations and pairs of bilinear forms, (1924). Proc. Intern. Congr. Math. Toronto, 1, 361–363.

418. Fantappiè, L.: Le calcul des matrices. Comptes Rendus 186, 619–621.

419. Giorgi, G.: Sulle funzioni delle matrici. R. Accad. Lincei Rend. (6) 7, 178–184.

420. Giorgi, G.: Nuove osservazioni sulle funzioni delle matrici. R. Accad. Lincei Rend. (6) 8, 3–8.

421. Hensel, K.: Über den Zusammenanhang zwischen den Systemen und ihren Determinanten. Crelle 159, 246–254.

422. Kowalewski, G.: Über lineare Differentialsysteme mit konstanten Koefficienten. Leipz. Ber. 80, 359–366.

423. MacDuffee, C. C.: A correspondence between matrices and quadratic ideals, (1927). Annals of Math. 29, 199–214.

424. Martis, S.: Ricerca di un'espressione razionale per le potenze di una matrice di secondo ordine. R. Accad. Lincei Rend. (6) 8, 130–133.

425. Martis, S.: Sugli esponenziali delle matrici di secondo ordine e loro applicazione alla teoria dei gruppi. R. Accad. Lincei Rend. (6) 8, 276–280.

426. Martis, S.: Calcolo del logaritmo di una matrice di secondo ordine, $\cdots$. R. Accad. Lincei Rend. (6) 8, 474–480.

427. Ory, H.: Sur l'équation $x^n = a$ ou a est un tableau carré du deuxième ordre. Comptes Rendus 187, 1012–1014.

428. Polya, G.: Über die Funktionalgleichung der Exponentialfunktionen im Matrizenkalkül. Berlin Sitzb. 1928, 96–99.

429. Porcu-Tortrini, E.: Calcolo delle funzioni qualunque di matrici di second' ordine. R. Accad. Lincei Rend. (6) 7, 206–208.

430. Roth, W. E.: A solution of the matrix equation $P(X) = A$, (1927). Trans. Amer. Math. Soc. 30, 579–596.

431. Schrukta, L.: Ein Beweis des Hauptsatzes der Theorie der Matrizes, (1924). Monatsh. f. Math. u. Phys. 35, 83–86.

432. Toscano, L.: Determinanti involutori, (1927). Rend. R. Ist. Lombardo (2) 61, 187–195.

433. Toscano, L.: Equazioni reciproche a matrice. R. Accad. Lincei Rend. (6) 8, 664–669.

434. Turnbull, H. W.: The theory of determinants, matrices, and invariants. (Dover)

435. Turnbull, H. W.: Non-commutative algebra, (1927). Math. Gaz. 14, 12–22.

436. Turnbull, H. W.: On differentiating a matrix, (1927). Proc. Edinb. Math. Soc. (2) 1, 111–128.

437. Vaidyanathaswamy, R.: Integer roots of the unit matrix. Journ. Lond. Math. Soc. 3, 121–124.

438. Vaidyanathaswamy, R.: On the possible periods of integer-matrices. Journ. Lond. Math. Soc. 3, 268–272.

439. Voghera, G.: Sulla forma canonica delle matrici, (1927). Boll. Un. Mat. Ital. 7, 32–34.

440. Weyl, H.: Gruppentheorie und Quantenmechanik. Leipzig.

1929

441. Beckenbach, E. F.: An inequality for definite hermitian determinants. Bull. Amer Math. Soc. 35, 325–329.

442. Cherubino, S.: Un'applicazione del calculo di matrici alla teoria delle forme quadratiche. Rend. R. Ist. Lomb. (2) 62, 505–514.

443. Cherubino, S.: Alcune osservazioni sulle matrici rettangolari e lora utilità per la decomposizione di una forma quadratica in somme di quadrati. Rend. R. Ist. Lomb. (2) 62, 623–649.

444. MacDuffee, C. C.: An introduction to the theory of ideals in linear associative algebras, (1927). Trans. Amer. Math. Soc. 31, 71–90.

445. MacDuffee, C. C.: On the independence of the first and second matrices of an algebra. Bull. Amer. Math. Soc. 35, 344–349.

446. Martis, S.: Ricerca di un'espressione razionale per le potenze di una matrice del terz'ordine. R. Accad. Lincei Rend. (6) 9, 206–213.

447. Pillai, S. S.: A theorem concerning the primitive periods of integer matrices. Journ. Lond. Math. Soc. 4, 250–251.

448. Roth, W. E.: A convenient check on the accuracy of the product of two matrices. Amer. Math. Monthly 36, 37–38.

449. Schlesinger, L.: Über den Logarithmus einer Matrix. Crelle 161, 199–200.

450. Scorza, G.: Sulle matrici di Riemann. R. Accad. Lincei Rend (6) 9, 253–258.

451. Shoda, K.: Über die mit einer Matrix vertauschbaren Matrizen. Math. Zeits. 29, 696–712.

452. Toscano, L.: Equazioni a matrice con radici reali. Tôhoku Math. Journ. 32, 27–31.

453. Tsuji, M.: On the roots of the characteristic equation of a certain matrix. Proc. Imp. Acad. Tokyo 5, 111–112.

454. Wintner, A.: Spektraltheorie der unendlichen Matrizen. Leipzig.

1930

455. Albert, A. A.: The non-existence of pure Riemann matrices with normal multiplication algebras of order sixteen, (1929). Annals of Math. 31, 375–380.

456. Albert, A. A.: On the structure of pure Riemann matrices with non-commutative multiplication algebras. Proc. Amer. Nat. Acad. Sci. 16, 308–312.

457. Albert, A. A.: On direct products, cyclic division algebras, and pure Riemann matrices. Proc. Amer. Nat. Acad. Sci. 16, 313–315.

458. Amante, S.: Risoluzione, nel campo delle matrici complesse, di una qualsiasi equazione analitica a coefficienti numerici. R. Accad. Lincei Rend. (6) 12, 290–296.

459. Bell, E. T.: A type of commutative matrices. Bull. Calcutta Math. Soc. 22, 53–60.

460. Born, M. and P. Jordan. Elementare Quantenmechanik Berlin.

461. Browne, E. T.: On the separation property of the roots of the secular equation. Amer. Journ. Math. 52, 843–850.

462. Browne, E. T.: The characteristic roots of a matrix. Bull. Amer. Math. Soc. 36, 705–710.

463. Dienes, P.: The exponential function in linear algebras. Quart. Journ. (2) 1, 300–309.

464. Ingraham, M. H.: An elementary theorem on matrices. Bull. Amer. Math. Soc. 36, 673–674.

465. Laura, E.: La teoria delle matrici e il metodo dell'n·edro mobile. Rend. Sem. Mat. Padova 1, 85–109.

466. Loewy, A. and R. Brauer. Über einen Satz für unitäre Matrizen. Tôhoku Math. Journ. 32, 44–49.

467. Mehmke, R.: Praktische Lösung der Grundaufgaben über Determinanten, Matrizen und lineare Transformationen. Math. Ann. 103, 300–318.

468. Mehmke, R.: Berichtigung, das Aufstellung der identischen Gleichung einer Matrix betreffend. Math. Ann. 104, 167–168.

469. Muir, T.: The literature of Cayleyan matrices. Trans. Roy. Soc. So. Africa 18, 219–225.

470. Nowlan, F. S.: On the direct product of a division algebra and a total matric algebra, (1927). Bull. Amer. Math. Soc. 36, 265–268.

471. Perron, O.: Über eine Matrixtransformation. Math. Zeits. 32, 465–473.

472. Rasch, G.: Om systemer af lineaere differentialligninger. Swende Skandin. Mat. Kongr., 1929, 117–126.

473. Rasch, G.: Om matrixregning og dens anvendelse paa differens- og differentialligninger. Diss. Copenhagen.

474. Roth, W. E.: On the unilateral equation in matrices, (1929). Trans. Amer. Math. Soc. 32, 61–80.

475. Tocchi, L.: Sulla caratteristica delle matrici. Giorn. di Mat. 68, 201–217.

476. Wellstein, J.: Über symmetrische, alternierende und orthogonale Normalformen von Matrizen. Crelle 163, 166–182.

1931

477. Aitken, A. C.: Further studies in algebraic equations and matrices. Proc. Roy. Soc. Edinb. 51, 80–90.

478. Albert, A. A.: The structure of matrices with any normal division algebra of multiplications, (1930). Annals of Math. 32, 131–148.

479. Cecioni, F.: Sull'equazione fra matrici $AX = XA$. Ann. Univ. Toscane 14, fasc. 2, 1–49.

480. Finan, E. J.: A determination of the domains of integrity of the complete rational matric algebra of order 4. Amer. Journ. Math. 53, 920–928.

481. Franklin, P.: Algebraic matric equations. Journ. Math. Phys. 10, 289–314.

482. Galanti, G.: Algoritmi di calcolo motoriale. Accad. Lincei Rend. (6) 13, 861–866.

483. Hasse, H.: Über p-adische Schiefkörper und ihre Bedeutung für die Arithmetik hyperkomplexer Zahlsysteme. Math. Ann. 104, 495–534.

484. Klein, F.: Über rechteckige Matrizen, bei denen die Determinanten maximaler Reihenanzahl teilerfremd zu einem Modul sind. Jbr. Deutsch. Math. Ver. 40, 233–238.

485. Kravcuk, M. and A. Smogorzevskij: Sur les transformations unitaires et orthogonales. Journ. Cycle Math. 1, 3–41.

486. Kravcuk, M.: Sur une inegalité. Bull. Acad. Sci. Ukraine 1, 96–101.

487. Littlewood, D. E.: Identical relations satisfied in an algebra, (1930). Proc. Lond. Math. Soc. (2) 32, 312–320.

488. MacDuffee, C. C.: The discriminant matrices of a linear associative algebra, (1930). Annals of Math. 32, 60–66.

489. MacDuffee, C. C.: The discriminant matrix of a semi-simple algebra, (1930). Trans. Amer. Math. Soc. 33, 425–432.

490. Murnaghan, F. D.: On the representation of a Lorentz transformation by means of two-rowed matrices. Amer. Math. Monthly 38, 504–511.

491. Murnaghan, F. D., and A. Wintner: A canonical form for real matrices under orthogonal transformations. Proc. Amer. Nat. Acad. Sci. 17, 417–420.

492. Nowlan, F. S.: A note on primitive idempotent elements of a total matric algebra, (1927). Bull. Amer. Math. Soc. 37, 854–856.

493. Romanovsky, V.: Sur les zéros des matrices stocastiques. Comptes Rendus 192, 266–269.

494. Schlesinger, L.: Neue Grundlagen für einen Infinitesimalkalkül der Matrizen, (1929). Math. Zeits. 33, 33–61.

495. Spampinato, N.: Algebre elementari, teoria delle semialgebre e cicli pseudoriemanniani. Note Esercit. Mat. 6, 107–245.

496. Takahashi, S.: On the roots of the characteristic equation of a certain matrix. Proc. Imp. Acad. Tokyo 7, 241–243.

497. Thomas, J. M.: Matrices of integers ordering derivatives. Trans. Amer. Math. Soc. 33, 389–410.

498. Thurston, H. S.: On the characteristic equations of products of square matrices. Amer. Math. Monthly 38, 322–324.

499. Thurston, H. S.: The characteristic equations of the adjoint and the inverse of a matrix. Amer. Math. Monthly 38, 448–449.

500. Turnbull, H. W.: The invariant theory of a general bilinear form, (1930). Proc. Lond. Math. Soc. (2) 33, 1–21.

501. Turnbull, H. W.: Matrix differentiation of the characteristic function. Proc. Edinb. Math. Soc. (2) 2, 256–264.

502. Ward, M.: The algebra of recurring series, (1930). Annals of Math. 32, 1–9.

503. Wedderburn, J. H. M.: Non-commutative domains of integrity. Crelle 167, 129–141.

504. Williamson, J.: Bazin's matrix and other allied matrices. Proc. Edinb. Math. Soc. (2) 2, 240–251.

505. Williamson, J.: The latent roots of a matrix of special type. Bull. Amer. Math. Soc. 37, 585–590.

506. Wintner, A., and F. D. Murnaghan: On a polar representation of non-singular matrices. Proc. Amer. Nat. Acad. Sci. 17, 676–678.

1932

507. Amato, V.: Sulla risoluzione di equazioni nel campo delle matrici complesse. Boll. Accad. Gioenia, Catania, Ser. (2), fasc. 62, 50–55.

508. Bottema, O.: Over machtreekseen van matrices. Nieuw Arch. Wisk. 17, 114–118.

509. Bush, L. E.: Note on the discriminant matrix of an algebra, (1931). Bull. Amer. Math. Soc. 38, 49–51.

510. Cipolla, M.: Sulle matrici espressioni analitiche di un'altra. Rend. Circ. Mat. Palermo 56, 144–154.

511. Eddington, A. S.: On sets of anti-commuting matrices, (1931). Journ. Lond. Math. Soc. 7, 58–68.

512. Menge, W. O.: On the rank of the product of certain square matrices, (1930). Bull. Amer. Math. Soc. 38, 88–94.
513. Mitchell, A. K.: On a matrix differential operator. Bull. Amer. Math. Soc. 38, 251–254.
514. Murnaghan, F. D.: On the unitary invariants of a square matrix. Proc. Amer. Nat. Acad. Sci. 185–189.
515. Mursi-Ahmed, M.: On the composition of simultaneous differential systems of the first order. Proc. Edinb. Math. Soc. (2) 3, 128–131.
516. Newman, M. H. A.: Note on an algebraic theorem of Eddington. Journ. Lond. Math. Soc. 7, 93–99.
517. Pierce, T. A.: The practical evaluation of resultants. Amer. Math. Monthly 39, 161–162.
518. Rutherford, D. E.: On the solution of the matrix equation $AX + XB = C$. Akad. Wet. Amsterdam Proc. 35, 54–59.
519. Rutherford, D. E.: On the canonical form of a rational integral function of a matrix. Proc. Edinb. Math. Soc. (2) 3, 135–143.
520. Rutherford, D. E.: On the rational commutant of a square matrix. Proc. Akad. Wet. Amsterdam 35, 870–875.
521. Schlesinger, L.: Weitere Beiträge zum Infinitesimalkalkül der Matrizen, (1931). Math. Zeits. 35, 485–501.
522. Schreier, O., and E. Sperner: Vorlesungen über Matrizen. Leipzig.
523. Smohorshevsky, A.: Sur les matrices unitaires du type de circulants. Journ. Cycle Math. 2, 89–90.
524. Takahashi, S.: Zum verallgemeinerten Infinitesimalkalkül der Matrizen. Jap. Journ. Math. 9, 27–46.
525. Turnbull, H. W., and A. C. Aitken: An introduction to the theory of canonical matrices. (Dover Reprint)
526. Weitzenböck, R.: Über die Matrixgleichung $AX + XB = C$. Proc. Akad. Wet. Amsterdam 35, 60–61.
527. Weitzenböck, R.: Über die Matrixgleichung $X^2 = A$. Proc. Akad. Wet. Amsterdam 35, 157–161.
528. Weitzenböck, R.: Über die Matrixgleichung $XX' = A$. Proc. Akad. Wet. Amsterdam 35, 328–330.
529. Williamson, J.: The product of a circulant matrix and a special diagonal matrix. Amer. Math. Monthly 39, 280–285.
530. Wilson, R.: Eliminants of the characteristic equations. Proc. Lond. Math. Soc. (2) 33, 517–524.

1933

531. Eddington, A. S.: On sets of anticommuting matrices. II. The factorization of E-numbers. Journ. Lond. Math. Soc. 8, 142–152.
532. Hensel, K.: Über die Ausführbarkeit der elementaren Rechenoperationen in Ringen von Systemen. Crelle 169, 67–70.
533. MacDuffee, C. C.: Matrices with elements in a principal ideal ring. Bull. Amer. Math. Soc. 39, 564–584.
534. MacDuffee, C. C.: The theory of matrices. Berlin.
535. Menge, W. O.: Construction of transformations to canonical forms. Amer. Journ. Math. 55, 671–682.
536. Milne-Thomson, L. M.: A matrix representation of ascending and descending continued fractions, (1932). Proc. Edinb. Math. Soc. (2) 3, 189–200.
537. Röseler, H.: Normalformen von Matrizen gegenüber unitären Transformationen. Diss. Darmstadt.

538. ROMANOVSKY, V.: Un théorème sur les zéros des matrices non négative. Bull. Soc. Math. France 61, 213–219.

539. ROTH, W. E.: On the equation $P(A, X) = 0$ in matrices. Trans. Amer. Math. Soc. 35, 689–708.

540. RUTHERFORD, D. E.: On the condition that two Zehfuss matrices be equal. Bull. Amer. Math. Soc. 39, 801–808.

541. RUTHERFORD, D. E.: On the rational solution of the matrix equation $SX = XT$. Proc. Akad. Wet. Amsterdam 36, 432–442.

542. SOKOLNIKOFF, E. S.: Matrices conjugate to a given matrix with respect to its minimum equation. Amer. Journ. Math. 55, 167–180.

543. TURNBULL, H. W.: Diagonal matrices. Proc. Cambr. Phil. Soc. 29, 347–372.

544. TURNBULL, H. W.: Matrices and continued fractions, II. Proc. Roy. Soc. Edinb. 53, 208–219.

545. WEGNER, U.: Über die Frobeniusschen Kovarianten. Monatsh. Math. Phys. 40, 201–208.

546. WEGNER, U. AND J. WELLSTEIN: Bemerkungen zur Transformation von Komplexen symmetrischen Matrizen. Monatsh. Math. Phys. 40, 319–322.

547. WILLIAMSON, J.: The expansion of determinants of composite order. Amer. Math. Monthly 40, 65–69.

548. WILLIAMSON, J.: Sets of semi-commutative matrices, (1932). Proc. Edinb. Math. Soc. (2) 3, 179–188.

549. WILLIAMSON, J.: Matrices whose sth compounds are equal. Bull. Amer. Math. Soc. 39, 108–111.

INDEX TO BIBLIOGRAPHY

INDEX

DOVER PHOENIX EDITIONS

An Introduction to the Approximation of Functions, Theodore J. Rivlin. Unabridged republication of the 1969 edition. 160pp. 49554-X

An Essay on the Foundations of Geometry, Bertrand Russell. Unabridged republication of the 1897 edition. 224pp. 49555-8

Elements of Number Theory, I. M. Vinogradov. Unabridged republication of the first edition, 1954. 240pp. 49530-2

Asymptotic Expansions for Ordinary Differential Equations, Wolfgang Wasow. Unabridged republication of the 1976 corrected, slightly enlarged reprint of the original 1965 edition. 384pp. 49518-3

Physics

Semiconductor Statistics, J. S. Blakemore. Unabridged, corrected, and slightly enlarged republication of the 1962 edition. 141 illustrations. xviii+318pp. 49502-7

Wave Propagation in Periodic Structures, L. Brillouin. Unabridged republication of the 1946 edition. 131 illustrations. 272pp. 49556-6

The Conceptual Foundations of the Statistical Approach in Mechanics, Paul and Tatiana Ehrenfest. Unabridged republication of the 1959 edition. 128pp. 49504-3

The Analytical Theory of Heat, Joseph Fourier. Unabridged republication of the 1878 edition. 20 figures. 496pp. 49531-0

States of Matter, David L. Goodstein. Unabridged republication of the 1975 edition. 154 figures. 4 tables. 512pp. 49506-X

The Principles of Mechanics, Heinrich Hertz. Unabridged republication of the 1900 edition. 320pp. 49557-4

Thermodynamics of Small Systems, Terrell L. Hill. Unabridged and corrected republication in one volume of the two-volume edition published in 1963–1964. 32 illustrations. 408pp. 6½ x 9¼. 49509-4

Theoretical Physics, A. S. Kompaneyets. Unabridged republication of the 1961 edition. 56 figures. 592pp. 49532-9

Quantum Mechanics, H. A. Kramers. Unabridged republication of the 1957 edition. 14 figures. 512pp. 49533-7

The Theory of Electrons, H. A. Lorentz. Unabridged reproduction of the 1915 edition. 9 figures. 352pp. 49558-2

The Principles of Physical Optics, Ernst Mach. Unabridged republication of the 1926 edition. 279 figures. 10 portraits. 336pp. 49559-0

The Scientific Papers of James Clerk Maxwell, James Clerk Maxwell. Unabridged republication of the 1890 edition. 197 figures. 39 tables. Total of 1,456pp. *Volume I* (640pp.) 49560-4; *Volume II* (816pp.) 49561-2

Vectors and Tensors in Crystallography, Donald E. Sands. Unabridged and corrected republication of the 1982 edition. xviii+228pp. 49516-7

Principles of Mechanics and Dynamics, Sir William (Lord Kelvin) Thompson and Peter Guthrie Tait. Unabridged republication of the 1912 edition. 168 diagrams. Total of 1,088pp. *Volume I* (528pp.) 49562-0; *Volume II* (560pp.) 49563-9

Treatise on Irreversible and Statistical Thermophysics: An Introduction to Nonclassical Thermodynamics, Wolfgang Yourgrau, Alwyn van der Merwe, and Gough Raw. Unabridged, corrected republication of the 1966 edition. xx+268pp. 49519-1

Engineering

Principles of Aeroelasticity, Raymond L. Bisplinghoff and Holt Ashley. Unabridged, corrected republication of the original 1962 edition. xi+527pp. 49500-0

Statics of Deformable Solids, Raymond L. Bisplinghoff, James W. Mar, and Theodore H. H. Pian. Unabridged and corrected Dover republication of the edition published in 1965. 376 illustrations. xii+322pp. 6½ x 9¼. 49501-9